The Biology, Ecology, and Management of Plant Pests

The Biology, Ecology, and Management of Plant Pests

Guest Editors

Lei Bian
Yu Gao
Suli Liu
Yuxin Zhou

MDPI

Basel • Beijing • Wuhan • Barcelona • Belgrade • Novi Sad • Cluj • Manchester

Guest Editors
Lei Bian
Tea Research Institute
Chinese Academy of
Agricultural Sciences
Hangzhou
China

Yu Gao
College of Plant Protection
Jilin Agricultural University
Changchun
China

Suli Liu
School of Mathematics
Jilin University
Changchun
China

Yuxin Zhou
College of Plant Protection
Jilin Agricultural University
Changchun
China

Editorial Office
MDPI AG
Grosspeteranlage 5
4052 Basel, Switzerland

This is a reprint of the Special Issue, published open access by the journal *Biology* (ISSN 2079-7737), freely accessible at: https://www.mdpi.com/journal/biology/special_issues/9P0464JJ0C.

For citation purposes, cite each article independently as indicated on the article page online and as indicated below:

Lastname, A.A.; Lastname, B.B. Article Title. *Journal Name* **Year**, *Volume Number*, Page Range.

ISBN 978-3-7258-8319-6 (Hbk)
ISBN 978-3-7258-8320-2 (PDF)
https://doi.org/10.3390/books978-3-7258-8320-2

Contents

Yu Gao, Suli Liu, Yuxin Zhou and Lei Bian
Advances in the Biology, Ecology, and Management of Plant Pests
Reprinted from: *Biology* **2026**, *15*, 901, https://doi.org/10.3390/biology15120901 1

Jiaqi Shi, Yuxin Zhou, Zhengxiao Du, Ruirui Li, Qi Wang, Yu Gao and Shusen Shi
Transcriptome Analysis and Identification of Chemosensory Genes in *Leguminivora glycinivorella*
Reprinted from: *Biology* **2026**, *15*, 505, https://doi.org/10.3390/biology15060505 6

Qingxuan Qiao, Yaqiong Chen, Jing Chen, Ting Chen, Huiting Feng, Yussuf Mohamed Salum, et al.
A Species-Specific *COI* PCR Approach for Discriminating Co-Occurring Thrips Species Using Crude DNA Extracts
Reprinted from: *Biology* **2026**, *15*, 171, https://doi.org/10.3390/biology15020171 32

Guochang Wang, Xing Ge, Dongbiao Lü, Ziyue Zhang, Li Wang, Saili Sun, et al.
Insecticidal Activity of *Angelica archangelica* Essential Oil and Transcriptomic Analysis of *Sitophilus zeamais* in Response to Oil Fumigation
Reprinted from: *Biology* **2025**, *14*, 1606, https://doi.org/10.3390/biology14111606 57

Yun He, Jiaru Zhang, Ziyang Hu, Yingxue Cao, Kevin H. Mayo, Duo Liu and Mingju E
Behavioral Responses of the Bumblebee *Bombus terrestris* to Volatile Compounds from Blueberries
Reprinted from: *Biology* **2025**, *14*, 1570, https://doi.org/10.3390/biology14111570 69

Lina Wang, Yuhua Situ, Jie Zhang, Kui Kang, Zhongjiu Xiao, Shaobo Wang, et al.
The Thermal Adaptability of *Sclerodermus guani* Xiao *et* Wu (Hymenoptera: Bethylidae), an Important Parasitoid of Long-Horned Beetles in China
Reprinted from: *Biology* **2025**, *14*, 1234, https://doi.org/10.3390/biology14091234 81

Mikaelly Batista da Silva, Ana Beatriz Medeiros, Antonia Isabelly Monteiro dos Anjos, João Vitor Ferreira Cavalcante, Bianca Cristiane Ferreira Santiago, Shênia Santos Monteiro, et al.
Changes in the Microbiota of the Scale Insect (*Diaspis echinocacti*, Bouché, 1833) in *Opuntia stricta* Cladodes: Taxonomic and Metagenomic Analysis as a Function of Infestation Levels
Reprinted from: *Biology* **2025**, *14*, 1233, https://doi.org/10.3390/biology14091233 93

Haiqin Yao, Haonan Xu, Jun Yang and Weihua Ma
Assessment of Potential Toxic Effects of RNAi-Based Transgenic Cotton on the Non-Target Predator *Harmonia axyridis*
Reprinted from: *Biology* **2025**, *14*, 1173, https://doi.org/10.3390/biology14091173 108

Chaowei Wang, Xingzhou Liu, Sylvestre T. O. Kelehoun, Kai Dong, Yueying Wang, Maozhu Yin, et al.
A Ubiquitous Volatile in Noctuid Larval Frass Attracts a Parasitoid Species
Reprinted from: *Biology* **2025**, *14*, 1007, https://doi.org/10.3390/biology14081007 122

biology

MDPI

Editorial

Advances in the Biology, Ecology, and Management of Plant Pests

Yu Gao [1], Suli Liu [2], Yuxin Zhou [1] and Lei Bian [3,*]

1 College of Plant Protection, Jilin Agricultural University, Changchun 130118, China; gaothrips@jlau.edu.cn (Y.G.)
2 School of Mathematics, Jilin University, Changchun 130012, China
3 Tea Research Institute, Chinese Academy of Agricultural Sciences, Hangzhou 310008, China
* Correspondence: bianlei@tricaas.com

1. Introduction

Plant pests pose a severe threat to global agricultural productivity and ecosystem stability [1,2]. Each year, crop losses caused by pests amount to hundreds of billions of dollars, gravely endangering food security and farmers' livelihoods [3]. Meanwhile, although traditional chemical control methods are effective in the short term, they face increasingly serious challenges: The rapid evolution of pest resistance leads to escalating pesticide use with diminishing returns and chemical residues contaminate soil and water sources, damaging the ecological environment [4]; moreover, broad-spectrum insecticides inflict fatal harm on beneficial non-target organisms such as bees and ladybugs, disrupting ecological balance [5]. Confronted with these dilemmas, the development of green and sustainable pest management strategies has become an urgent global agricultural necessity [6]. Environmentally friendly technologies, such as biological control [7] and pest-resistant crop breeding [8], are emerging as critical pathways to replace traditional chemical control and achieve sustainable agriculture.

The nine contributions (eight articles and one editorial) in this Special Issue, 'Biology, Ecology, and Management of Plant Pests', showcase cutting-edge advances in plant pest research across three integrated dimensions: biology, ecology, and the management of plant pests. Biology encompasses the molecular basis of chemosensory systems and the physiological toxicity and molecular response mechanisms of plant essential oils. Ecology addresses multi-trophic interactions and environmental adaptation, including the thermal plasticity of parasitoid wasps, plant–pollinator volatile signaling, microbiome dynamics of scale insects, and microbe-mediated kairomone systems between noctuid hosts and their parasitoids. Management focuses on practical tools and strategies, spanning species-specific molecular diagnostics for thrips, and the environmental risk assessment of RNAi-based transgenic cotton for non-target predators.

2. Biology of Plant Pests

The following two articles, from the perspectives of the molecular basis of chemosensory systems and the physiological toxicity and molecular response mechanisms of plant essential oils, respectively, reflect the fundamental research on physiological mechanisms, molecular foundations, and other aspects of pests within the framework of the biology of plant pests.

Shi et al. conducted transcriptome analysis across multiple adult tissues of the soybean pod borer *Leguminivora glycinivorella* to identify chemosensory genes [9]. A comprehensive repertoire was annotated, including 76 odorant receptors (ORs), 15 gustatory receptors (GRs), 18 ionotropic receptors (IRs), 52 odorant-binding proteins (OBPs), 18 chemosensory proteins, and four sensory neuron membrane proteins (SNMPs). Notably, canonical bitter GRs and specific IR lineages (IR100/IR85a) were absent, potentially reflecting the adaptation to specialized soybean-feeding. Pronounced sexual and tissue dimorphism was observed: male antennae were enriched for putative pheromone receptors and *LglySNMP1*, while female-biased OBPs and ORs suggested roles in host location and oviposition. This study provides a valuable expression atlas for developing behavior-based green pest management strategies.

Wang et al. (2025) evaluated the insecticidal activity of *Angelica archangelica* root essential oil (AAEO) against the maize weevil *Sitophilus zeamais* through gas chromatography–mass spectrometry (GC-MS) analysis, fumigation toxicity assays, enzyme activity measurements, and transcriptome sequencing [10]. GC-MS identified 35 components, with δ-3-Carene, Limonene, and α-Pinene as major constituents. The LC_{50} values were 164.38, 132.62, and 90.35 mg/L air at 24, 48, and 72 h, respectively. AAEO significantly inhibited acetylcholinesterase, glutathione S-transferase, and carboxylesterase activities. RNA-seq revealed 3718 significantly differentially expressed genes, with qRT-PCR confirming the reliability of the transcriptome data. This study provides a scientific basis for developing effective plant essential oil-based bioinsecticides against stored-grain pests.

3. Ecology of Plant Pests

The following four articles from the perspectives of temperature-driven natural enemy population dynamics and climate adaptation, multi-trophic chemical information transfer and microbe-mediated interaction networks, plant–pollinator chemical communication and community function optimization, and pest microbiome community structure and host–plant interactions, respectively, reflect the core essence of ecological-level research in the ecology of plant pests, encompassing pest–environment interactions, population dynamics, community structure, food-web relationships, and climate adaptation, while also laying a theoretical foundation for the development of green pest control technologies.

The study by Wang et al. (2025) systematically evaluated the effects of temperature (21–33 °C) on the parasitism capacity and reproductive fitness of *Sclerodermus guani* [11]. Results showed that temperature significantly affected female wasp parasitoid behavior, the pre-oviposition period, and offspring developmental duration: as the temperature increased, host-seeking and stinging behaviors became more active, and the pre-oviposition period shortened from 31.3 days (21 °C) to 4.1 days (33 °C). Reproductive fitness peaked at 27–30 °C, with both low and high temperatures being detrimental; offspring numbers decreased by 30% at 33 °C compared to the maximum. The developmental threshold temperatures for the egg, larval, and pupal stages were calculated as 10.19 °C, 7.73 °C, and 15.57 °C, respectively. This study provided biological data above 30 °C for the first time, offering important guidance for breeding heat-tolerant strains and optimizing indoor rearing and field release strategies under climate change.

The study by Wang et al. (2025) investigated the relative attractiveness of the host larval body and frass volatiles to the parasitoid *Microplitis mediator*, using two noctuid pests (*Helicoverpa armigera* and *Spodoptera frugiperda*) as hosts [12]. Through four-arm olfactometer bioassays, GC-MS analysis, and gas chromatography–electroantennographic detection (GC–EAD), larval frass extracts were found to be significantly more attractive to parasitoids than body extracts, and this attractiveness was independent of host food types (maize, cotton, soybean leaves, or artificial diet). Ethyl palmitate was identified as the key attractant compound in frass, eliciting strong antennal responses and behavioral attraction in parasitoids. Antibiotic treatment experiments indicated that ethyl palmitate

https://doi.org/10.3390/biology15120901

production was associated with gut bacterial metabolism. This study revealed for the first time that a ubiquitous gut microbe-derived metabolite in noctuid larval frass serves as a critical infochemical for parasitoid host location, providing novel insights for enhancing biological control using volatile attractants.

The study by da Silva et al. investigated changes in the microbiota of the armored scale insect *Diaspis echinocacti* infesting *Opuntia stricta cladodes* across low, intermediate, and high infestation levels using 16S rRNA amplicon sequencing [13]. The obligate endosymbiont *Candidatus Uzinura* diaspidicola dominated all samples, indicating a resilient core microbiome. Shannon diversity showed a declining trend from low to high infestation, and PICRUSt2 predicted a contraction of metabolic breadth at higher infestations with convergence on energy and amino acid biosynthesis pathways. This study provides baseline data on scale insect microbiota dynamics and their potential role in host–plant interactions, guiding future microbiome-based strategies for sustainable cactus protection.

The study of He et al. (2025) identified 32 volatile compounds from Changbai Mountain blueberry flowers using solid-phase microextraction (SPME) combined with GC-MS, with linalool and styrene as the primary components [14]. Through electroantennography and Y-tube olfactometer behavioral assays, six compounds with significant electrophysiological activity in bumblebees were screened (benzaldehyde, phenylpropylaldehyde, citral, linalool, α-terpineol, and geraniol). The study found that low concentrations of geraniol, linalool, and α-terpineol attracted bumblebees, while high concentrations of benzaldehyde, phenylpropylaldehyde, and citral showed repellent effects. This research systematically elucidated the chemical interaction mechanism between blueberry floral volatiles and bumblebees for the first time, providing a theoretical basis for developing natural pollination attractants to improve blueberry yield and quality.

4. Management of Plant Pests

The following two articles from the perspectives of molecular monitoring technology development and application and the ecological risk assessment and safety management of genetically modified crops, respectively, reflect the core essence of applied research in Integrated Pest Management, encompassing pest monitoring, control technology development, risk assessment, and comprehensive management strategies. They provide important technical support and theoretical foundations for accurate pest identification, early warning, and sustainable management.

Qiao et al. developed a species-specific *COI* PCR approach for discriminating four co-occurring thrips species (*Frankliniella intonsa*, *F. occidentalis*, *Megalurothrips usitatus*, and *Thrips hawaiiensis*) using crude DNA extracts [15]. By designing four species-specific primer pairs targeting polymorphism-rich regions of the mitochondrial *COI* gene with diagnostic 3′-terminal nucleotides, combined with a PBS-based short-turnaround DNA extraction workflow, the assay achieved a practical detection limit of 1 ng per reaction and maintained reproducible amplification under mixed-species backgrounds. PCR inhibition from crude lysates was effectively alleviated by fivefold dilution. This framework provides a field-adaptable tool for thrips species identification and invasive species monitoring, with potential for integration with visual nucleic acid detection platforms.

The study of Yao et al. (2025) systematically assessed the potential ecological risks of dsAsFAR transgenic cotton targeting *Adelphocoris suturalis* on the non-target predatory lady beetle *Harmonia axyridis* [16]. Through two exposure pathways—direct artificial diet feeding and trophic transfer—combined with transcriptome sequencing and qRT-PCR validation, results showed that dsAsFAR had no significant adverse effects on lady beetle growth, development, survival, predation ability, fecundity, or FAR homologous gene expression. Although a few off-target differentially expressed genes were detected,

https://doi.org/10.3390/biology15120901

Shannon entropy analysis indicated that the overall transcriptome homeostasis was not disturbed. This study established a comprehensive risk assessment framework including exposure pathway evaluation, worst-case feeding trials, off-target effect analysis, and transcriptome stability assessment, providing an important reference for the biosafety evaluation of RNAi transgenic crops.

This Special Issue covers the biology, ecology, and management of plant pests, spanning chemosensory mechanisms, essential oil insecticidal activity, parasitoid thermal plasticity, microbe-mediated kairomones, scale insect microbiome dynamics, blueberry-pollinator signaling, thrips molecular diagnostics, and RNAi cotton risk assessment, offering key foundations for green pest management. Future work should harness multi-omics and AI to map pest–plant–enemy–microbe networks and pinpoint new control targets [17–19]. Sharper predictive models for pest adaptation and resistance under climate change, plus the field-scale deployment of RNAi biopesticides, pheromone traps, and microbial agents, will prove vital [20,21]. Building an integrated management system rooted in ecological regulation and smart monitoring can thus secure both agricultural sustainability and ecological safety.

Author Contributions: Conceptualization, Y.G. and L.B.; writing—original draft preparation, Y.G., Y.Z., S.L. and L.B.; writing—review and editing, Y.G., Y.Z., S.L. and L.B. All authors have read and agreed to the published version of the manuscript.

Funding: This work was supported by the Earmarked Fund for China Agriculture Research System of MOF and MARA (Grant number CARS–04).

Data Availability Statement: The original contributions presented in this study are included in the list of contributions. Further inquiries can be directed to the corresponding authors.

Acknowledgments: We extend our sincere gratitude to all contributing authors, the dedicated reviewers, and the editorial team at *Biology* for their invaluable efforts in the success of this Special Issue.

Conflicts of Interest: The authors declare no conflicts of interest.

References

1. Oerke, E.-C. Crop losses to pests. *J. Agric. Sci.* **2006**, *144*, 31–43. [CrossRef]
2. Fenibo, E.O.; Ijoma, G.N.; Matambo, T. Biopesticides in sustainable agriculture: Current status and future prospects. In *New and Future Development in Biopesticide Research: Biotechnological Exploration*; Mandal, S.D., Ramkumar, G., Karthi, S., Jin, F., Eds.; Springer: Singapore, 2022. [CrossRef]
3. Savary, S.; Willocquet, L.; Pethybridge, S.J.; Esker, P.; McRoberts, N.; Nelson, A. The global burden of pathogens and pests on major food crops. *Nat. Ecol. Evol.* **2019**, *3*, 430–439. [CrossRef] [PubMed]
4. Pei, T.; Wang, L.; Zhao, Y.; Shi, S.; Gao, Y. Toxicity and efficacy of thirty insecticides against *Thrips flavus* in Northeast China: Laboratory, semifield, and field trials. *Insects* **2025**, *16*, 405. [CrossRef] [PubMed]
5. Razmkabir, M.; Youzi, S.; Adnani, F.; Mahmoudi, P.; Hatjina, F. Toxicity of neonicotinoid insecticides to honey bees: A network meta-analysis. *Apidologie* **2026**, *57*, 35. [CrossRef]
6. Shao, X.; Zhang, Q.; Li, L.; Tan, R.; Xu, K. Host plant apparency and push–pull strategies: A unified framework linking plant-mediated mechanisms for sustainable pest management. *Insects* **2026**, *17*, 543. [CrossRef]
7. Jaros-Tsoj, K.; Jaroszuk-Ściseł, J.; Oleńska, E.; Sugier, P.; Rineau, F.; Vassilev, A.; Vangronsveld, J.; Wójcik, M. Biostimulants for sustainable agriculture: Enhancing plant growth and stress resilience. *Appl. Sci.* **2026**, *16*, 4685. [CrossRef]
8. Zhang, X.; Yang, J.; Chen, Y.; Wang, M.; Zhang, X.; Shen, M.; Zhang, S.; Wang, Z.; Wang, X. Breeding and molecular characterization of insect-resistant transgenic cotton. *Plants* **2026**, *15*, 1551. [CrossRef] [PubMed]
9. Shi, J.; Zhou, Y.; Du, Z.; Li, R.; Wang, Q.; Gao, Y.; Shi, S. Transcriptome analysis and identification of chemosensory genes in *Leguminivora glycinivorella*. *Biology* **2026**, *15*, 505. [CrossRef] [PubMed]
10. Wang, G.; Ge, X.; Lü, D.; Zhang, Z.; Wang, L.; Sun, S.; Jia, X.; Zhang, B.; Kuanysh, K.; Dani, S.; et al. Insecticidal activity of *Angelica archangelica* essential oil and transcriptomic analysis of *Sitophilus zeamais* in response to oil fumigation. *Biology* **2025**, *14*, 1606. [CrossRef] [PubMed]

 https://doi.org/10.3390/biology15120901

11. Wang, L.; Situ, Y.; Zhang, J.; Kang, K.; Xiao, Z.; Wang, S.; Wei, K.; Tang, Y. The thermal adaptability of *Sclerodermus guani* Xiao et Wu (Hymenoptera: Bethylidae), an important parasitoid of long-horned beetles in China. *Biology* **2025**, *14*, 1234. [CrossRef] [PubMed]
12. Wang, C.; Liu, X.; Kelehoun, S.T.O.; Dong, K.; Wang, Y.; Yin, M.; Li, J.; Gao, Y.; Xu, H. A ubiquitous volatile in noctuid larval frass attracts a parasitoid species. *Biology* **2025**, *14*, 1007. [CrossRef] [PubMed]
13. da Silva, M.B.; Medeiros, A.B.; dos Anjos, A.I.M.; Ferreira Cavalcante, J.V.; Santiago, B.C.F.; Monteiro, S.S.; Vital, A.C., Jr.; Dalmolin, R.J.S.; Lisboa, H.M.; Pasquali, M.A.d.B. Changes in the microbiota of the scale insect (*Diaspis echinocacti*, Bouché, 1833) in *Opuntia stricta* Cladodes: Taxonomic and metagenomic analysis as a function of infestation levels. *Biology* **2025**, *14*, 1233. [CrossRef] [PubMed]
14. He, Y.; Zhang, J.; Hu, Z.; Cao, Y.; Mayo, K.H.; Liu, D.; E, M. Behavioral responses of the bumblebee *Bombus terrestris* to volatile compounds from blueberries. *Biology* **2025**, *14*, 1570. [CrossRef] [PubMed]
15. Qiao, Q.; Chen, Y.; Chen, J.; Chen, T.; Feng, H.; Salum, Y.M.; Wang, H.; Tang, L.; Zhang, H.; Chen, Z.; et al. A species-specific *COI* PCR approach for discriminating co-occurring thrips species using crude DNA extracts. *Biology* **2026**, *15*, 171. [CrossRef] [PubMed]
16. Yao, H.; Xu, H.; Yang, J.; Ma, W. Assessment of potential toxic effects of RNAi-based transgenic cotton on the non-target predator *Harmonia axyridis*. *Biology* **2025**, *14*, 1173. [CrossRef] [PubMed]
17. Majić, I.; Ereš, H.; Plaščak, I.; Ozimec, S.; Rožac, V.; Sarajlić, A. Bioacoustic monitoring and AI applications in insect pest management. *Appl. Sci.* **2026**, *16*, 5176. [CrossRef]
18. Araújo, G.N.; Viteri Jumbo, L.O.; Silva, P.B.; Souza, L.B.; Silva, A.C.A.; Anhezini, L.; Santos, G.R.; Aguiar, R.W.S.; Oliveira, E.E.; Oliveira, J.M. Plant-based insecticides for controlling *Drosophila suzukii*: Opportunities and challenges for biorational nanoproducts. *Agrochemicals* **2026**, *5*, 8. [CrossRef]
19. Li, W.; Chen, X.; Zhou, Y.; Pei, T.; Liu, S.; Gao, Y. Dynamic predation model for controlling soybean aphids (*Aphis glycines*): A case study of simulated artificial release of ladybugs (*Harmonia axyridis*). *Agronomy* **2026**, *16*, 861. [CrossRef]
20. Rahman, M.T.; Bakibillah, A.S.M.; Hossain, A.; Ahasan, A.; Basher, M.N.; Oyshe, K.U.; Mariam, A. The role of AI-integrated drone systems in agricultural productivity and sustainable pest management. *AgriEngineering* **2026**, *8*, 142. [CrossRef]
21. Alves, A.d.C.; Silva, A.C.; Silva, A.T.d.; Sales, N.K.L.; Mamani, R.C.C.; Figueroa, L.B.P.; Gomes, E.H.; de Souza, D.C.T.; Marucci, R.C.; Soares, F.E.d.F. Compatibility between *Beauveria bassiana* and papain and their synergistic potential in the control of *Tenebrio molitor* (Coleoptera: Tenebrionidae). *Agrochemicals* **2026**, *5*, 2. [CrossRef]

https://doi.org/10.3390/biology15120901

biology

MDPI

Article

Transcriptome Analysis and Identification of Chemosensory Genes in *Leguminivora glycinivorella*

Jiaqi Shi [1], Yuxin Zhou [1,2,*], Zhengxiao Du [1], Ruirui Li [1], Qi Wang [1], Yu Gao [1] and Shusen Shi [1,*]

1 Key Laboratory of Soybean Disease and Pest Control, Ministry of Agriculture and Rural Affairs, College of Plant Protection, Jilin Agricultural University, Changchun 130118, China; shijiaqi0517@163.com (J.S.); wisdomking@foxmail.com (Z.D.); liruirui2026@163.com (R.L.); wq980828@126.com (Q.W.); gaothrips@jlau.edu.cn (Y.G.)

2 State Key Laboratory of Green Pesticide, Guizhou University, Guiyang 550025, China

* Correspondence: zhouyx@jlau.edu.cn (Y.Z.); sss-63@263.net (S.S.)

Simple Summary

The soybean pod borer is a serious soybean pest because its larvae feed inside pods and damage developing seeds, making conventional insecticide control difficult. To support the development of safer and more effective management strategies, we investigated the genes that may help this insect detect chemical cues from host plants, mates, and its surroundings. By analyzing transcriptome data from multiple adult tissues, we identified 183 candidate genes related to chemical sensing. Many of these genes showed clear differences in expression among tissues and between females and males. In particular, genes associated with odor detection were mainly enriched in the antennae, the primary sensory organs for smelling. Male antennae showed high expression of several candidate genes that may be involved in pheromone detection, whereas some genes were more highly expressed in females and may contribute to host finding and egg-laying site selection. We also found that some expected sensory gene groups were absent from the dataset, which may be related to the insect's close adaptation to soybean. These results provide a foundation for future functional studies and may help support environmentally friendly pest control based on chemical communication.

Abstract

The soybean pod borer, *Leguminivora glycinivorella*, is a monophagous pest that threatens soybean production. Its larvae feed concealed within pods, which limits the efficacy of conventional insecticides. Elucidating its chemosensory system is therefore essential for developing green, behavior-based management strategies. Reference-based transcriptomics across multiple tissues of *L. glycinivorella* identified a comprehensive repertoire of chemosensory genes, including 76 odorant receptors (ORs), 15 gustatory receptors (GRs), 18 ionotropic receptors (IRs), 52 odorant-binding proteins (OBPs), 18 chemosensory proteins (CSPs), and 4 sensory neuron membrane proteins (SNMPs). Sequence and phylogenetic analyses characterized these candidates within the context of known insect chemosensory families. Notably, canonical bitter GRs and specific IR lineages (e.g., IR100/IR85a) were not detected in our dataset, potentially reflecting adaptation to the specialized soybean-feeding habit of this pest. Expression profiling further revealed pronounced sexual and tissue dimorphism: male antennae showed significant enrichment of putative pheromone receptors (PRs) and *LglySNMP1*, whereas several OBPs and ORs exhibited female-biased expression, suggesting roles in host location and oviposition. Additionally, the high expression of GR43a homologs points to fructose sensing, while the lack of detectable CO_2 receptor components (except *LglyGR2*) suggests atypical carbon dioxide perception mechanisms.

Collectively, this study provides a valuable expression atlas of chemosensory genes in *L. glycinivorella* and identifies sex-specific candidate genes for future functional validation and behavior-based pest management.

Keywords: *Leguminivora glycinivorella*; Olethreutidae; chemosensory genes; transcriptome

1. Introduction

The soybean pod borer, *L. glycinivorella* (Matsumura, 1898), a member of the family Olethreutidae (Lepidoptera), is a univoltine, highly host-specific pest that overwinters as mature larvae in the soil [1]. It is widely distributed across China, Japan, and other parts of East Asia, and ranks as one of the most destructive pests of soybean in northeastern and northern China [1]. Larvae bore into pods and feed on developing seeds, causing significant yield losses (typically 10–20%, up to >40% in severe outbreaks) and reducing grain quality and market value; infestation levels have been reported to increase in some regions/years [2]. Because larvae feed inside sealed pods, conventional chemical insecticides cannot effectively reach them, posing major challenges for control [3]. Consequently, research has increasingly shifted toward chemical signaling, and deciphering the molecular mechanisms underlying insect chemoperception has become an important avenue for developing novel green pest management strategies. Recent work has advanced our understanding of *L. glycinivorella*'s diapause regulation [4], transgenic resistance [5], microbial control [6], and biological control [7,8] and even identified new associated pests like *Orosius orientalis* (Matsumura, 1914) [9]. However, despite these efforts, the molecular basis of its olfactory system, which is central to host finding, mate location, and oviposition, remains uncharacterized. To date, no comprehensive study has identified or profiled its chemosensory receptor genes, leaving a critical gap in the development of behavior-based management tools.

In recent years, research on insect chemosensory systems has emerged as a focal point at the intersection of chemical ecology, behavioral biology, and sustainable pest control [10,11]. Insects rely on highly specialized chemosensory mechanisms to detect complex environmental cues—such as plant volatile organic compounds (VOCs), sex pheromones, oviposition stimulants, and defensive secondary metabolites [10]. This process is mediated by several conserved gene families: odorant receptors (ORs), gustatory receptors (GRs), ionotropic receptors (IRs), odorant-binding proteins (OBPs), chemosensory proteins (CSPs), and sensory neuron membrane proteins (SNMPs) [10–14].

Specifically, ORs are seven-transmembrane proteins that function as ligand-gated cation channels through heteromeric complex formation with the conserved co-receptor Orco. Ligand binding to plant terpenes or pheromone components triggers channel opening and the rapid activation of olfactory neurons [15,16]. GRs share a similar seven-transmembrane topology but are evolutionarily distinct from ORs—some evidence suggests that ORs may have originated from GRs. GRs primarily detect non-volatile compounds (e.g., sugars and bitter substances) via contact chemosensation and also mediate CO_2 perception; they are abundantly expressed in mouthparts and are crucial for host recognition and feeding decisions [17]. In Lepidoptera, IRs are classified into three subfamilies: antennal IRs (A-IRs), lepidopteran-specific IRs (LS-IRs), and divergent IRs (D-IRs) [18]. A-IRs, expressed in antennae, detect specific odorants such as acids and amines [19–23] and also contribute to thermosensation and hygrosensation [24–27]. D-IRs, found in gustatory tissues but not antennae, are implicated in taste reception [28–32], while LS-IR functions remain unclear. Functional IR complexes often require co-receptors IR8a or IR25a. For

 https://doi.org/10.3390/biology15060505

example, in *Bactrocera dorsalis* (Hendel, 1912), CRISPR/Cas9 knockout of IR8a impairs acetic acid detection, and calcium imaging shows that IR64a and IR75a (paired with IR76b or IR8a) act as candidate acid receptors, potentially complementing detection channels mediated by ORs/GRs [31,32].

Beyond membrane receptors, soluble carrier proteins—OBPs and CSPs—play essential roles in transporting hydrophobic odorants through aqueous sensillum lymph to receptors. OBPs typically contain six conserved cysteines and exhibit selectivity for odor molecules, accumulating in antennal lymph [33]. Although the functional necessity of OBPs has been questioned in some insect models, they are widely regarded as critically important for olfactory perception in Lepidoptera Odorant Reception in Insects: Functional and Evolutionary Perspectives. CSPs, with only four conserved cysteines, possess more compact, thermostable structures and broader expression—not only in olfactory organs but also in wings, legs, and hemolymph—suggesting roles beyond olfaction, such as in development or immunity [34]. SNMPs, CD36-family transmembrane glycoproteins localized on olfactory neuron membranes, enhance sensitivity to sex pheromones by interacting with pheromone receptors; loss of SNMP1 can severely disrupt mating behavior [35,36].

Notably, these chemosensory genes display pronounced tissue specificity and sexual dimorphism. For instance, in *Helicoverpa armigera* (Hübner, 1808), *HarmSNMP1* is highly expressed in antennae and is critical for pheromone detection, while *HarmGR9*—abdomen-enriched—regulates satiety and sugar intake [17]. In *Spodoptera littoralis* (Boisduval, 1833), SlitIR87a is antenna-specific, implicating it in odor sensing [31]; in *S. frugiperda*, *SfruCSP1/2* are highly expressed in larval cuticle and may serve as targets for insecticides like chlorfenapyr [37]. In *H. assulta* and *H. armigera*, mouthpart-biased CSP4 facilitates sugar feeding by reducing surface tension [36]. Moreover, in *Holotrichia parallela*, the pheromone receptor OR14 shows male-biased, circadian-rhythmic expression synchronized with its every-other-day pheromone release behavior [38]. Collectively, these findings reveal that chemosensory genes are deeply integrated into the behavioral regulatory networks governing the insect life cycle [35,39], and their sex- and tissue-biased expression provides key clues for dissecting behaviors like female oviposition preference or male pheromone tracking.

Despite extensive characterization in other major lepidopteran pests, the molecular landscape of the soybean pod borer's chemosensory system remains unexplored. Current knowledge is limited to the identification of its sex pheromones—primarily (E)-10-dodecen-1-yl acetate and (E,E)-8,10-dodecadien-1-yl acetate—and associated electrophysiological responses (e.g., EAG and GC–EAD) [9]. However, the molecular receptors and binding proteins involved in detecting these pheromones and host volatiles have not been systematically identified, especially those in the GR and IR families.

To address this gap, we performed transcriptome sequencing of multiple adult tissues of *L. glycinivorella*, including male and female antennae, heads without antennae, thoraces, abdomens, and legs. We comprehensively annotated all six core chemosensory gene families (ORs, GRs, IRs, OBPs, CSPs, and SNMPs), analyzed their phylogenetic relationships with homologs from other Lepidoptera, and integrated tissue-specific expression profiles to prioritize candidate genes potentially involved in (1) sex pheromone perception (e.g., ORs, OBPs, and SNMPs), (2) host volatile detection (e.g., ORs and IRs), and (3) contact-mediated oviposition behavior (e.g., GRs, IRs, and CSPs). This study fills a critical gap in the molecular understanding of *L. glycinivorella* chemoreception and provides a foundational resource and promising molecular targets for future functional validation and the development of chemosensory-based green control strategies, such as behavioral disruptors, pheromone synergists, or RNA-targeted biopesticides.

 https://doi.org/10.3390/biology15060505

2. Materials and Methods

2.1. Insect Source and Tissue Sampling

The *L. glycinivorella* colony used in this study was maintained in the insectary of the Economic Crop Pest Management and Control Team, Jilin Agricultural University. The colony was originally established from individuals collected at the Xidi experimental field in Changchun, Jilin Province, China (43.81061° N, 125.40368° E). Insects were reared for multiple generations under controlled laboratory conditions at 25 ± 1 °C, 65 ± 5% relative humidity, and a 16 h light:8 h dark photoperiod. Newly emerged adults were supplied with 10% (*v*/*v*) honey solution.

To obtain chemosensory-related transcripts from different adult tissues, eight tissue types were dissected from 3-day-old female and male adults, including female antennae (FA), male antennae (MA), female heads without antennae (FH), male heads without antennae (MH), mixed thoraces (T), mixed legs (L), female abdomens (FAb), and male abdomens (MAb). All dissections were performed on ice. Collected tissues were immediately frozen in liquid nitrogen and stored at −80 °C until RNA extraction.

2.2. RNA Extraction and Transcriptome Sequencing

Total RNA was extracted from each tissue using TRIzol reagent (Invitrogen, Thermo Fisher Scientific, Waltham, MA, USA). RNA purity and concentration were assessed using a NanoDrop 2000 spectrophotometer (Thermo Scientific, Thermo Fisher Scientific, Waltham, MA, USA), and RNA integrity was verified with an Agilent 2100 Bioanalyzer (Agilent Technologies, Santa Clara, CA, USA) or LabChip GX system (Revvity, Hopkinton, MA, USA). Qualified RNA samples (RIN ≥ 7.0) were sent to Biomarker Technologies (Beijing, China) for library construction and sequencing. Stranded cDNA libraries were prepared using oligo (dT) magnetic beads to enrich poly(A)+ mRNA. The mRNA was fragmented, reverse-transcribed into first- and second-strand cDNA, and then subjected to end repair, A-tailing, and adapter ligation. Libraries were size-selected using AMPure XP beads (Beckman Coulter, Brea, CA, USA), followed by PCR amplification. Library quality was assessed using a Qubit 3.0 fluorometer (Invitrogen, Thermo Fisher Scientific, Waltham, MA, USA), and libraries with concentrations ≥ 1 ng/μL were retained. Insert size distribution was validated on a Qsep400 system (BiOptic Inc., New Taipei City, Taiwan), and effective library concentration (>2 nM) was determined by qPCR. Qualified libraries were sequenced on an Illumina platform in paired-end 150 bp (PE150) mode.The raw transcriptome sequencing data generated in this study have been deposited in the NCBI Sequence Read Archive (SRA) under BioProject accession number PRJNA1434332. The corresponding BioSample and SRA run accession numbers are provided in Supplementary Table S1.

2.3. Transcriptome Assembly and Functional Annotation

Raw reads were quality-filtered to remove adapter-contaminated and low-quality reads, yielding clean data. Clean reads were aligned to the *L. glycinivorella* reference genome (*Leguminivora_glycinivorella*.GCF_023078275.1.genome.fa) using HISAT2 v2.0.4 [40]. Transcriptome reconstruction and expression quantification were performed using StringTie v2.2.1 [41], and gene expression levels were normalized as FPKM. Functional annotation was conducted using the BMKCloud (Biomarker Technologies) standard pipeline, including DIAMOND searches against Nr/Swiss-Prot/KEGG and related databases [42], GO assignment using InterProScan [43], and conserved domain identification using Pfam [44].

2.4. Identification and Structural Prediction of Chemosensory Genes

Six chemosensory gene families—OBP, CSP, OR, IR, GR, and SNMP—were identified from the annotated transcriptome. To improve annotation accuracy, homologous sequences

 https://doi.org/10.3390/biology15060505

from closely related lepidopterans were used as references, and candidate genes were confirmed by BLASTP searches against the NCBI database. Open reading frames (ORFs) were predicted using (NCBI, https://www.ncbi.nlm.nih.gov/orffinder/, accessed on 13 March 2026). Signal peptides were predicted using SignalP 5.0 (https://services.healthtech.dtu.dk/service.php?SignalP-5.0, accessed on 13 March 2026), and transmembrane domains were predicted using the TOPCONS web server (https://topcons.cbr.su.se/, accessed on 13 March 2026). Molecular weight and isoelectric point were calculated using the ExPASy Compute pI/Mw tool (https://web.expasy.org/compute_pi/, accessed on 13 March 2026). Amino acid sequence alignment and manual editing were performed using DNAMAN 8.0.

2.5. Phylogenetic Analysis

Amino acid sequences of *L. glycinivorella* chemosensory genes and homologous sequences from representative insect species were aligned using MAFFT v7.52 [45]. Poorly aligned and highly gapped regions were trimmed prior to phylogenetic reconstruction. Maximum-likelihood (ML) phylogenetic trees were constructed using IQ-TREE v1.6.12 [46]. The best-fit amino acid substitution model for each dataset was automatically selected using ModelFinder implemented in IQ-TREE [47], with JTT + F + R7 for ORs, VT + F + G4 for GRs, JTT + F + R8 for IRs, LG + R4 for OBPs, LG + F + R4 for CSPs, and LG + I + G4 for SNMPs. Node support was assessed using 10,000 ultrafast bootstrap replicates [48]. Because these chemosensory gene families are highly divergent and suitable outgroups were not available for reliable rooting, the ML phylogenetic trees were presented and interpreted as unrooted trees.

To further evaluate the robustness of the phylogenetic inference, Bayesian inference (BI) analyses were additionally performed based on CDS datasets. CDS sequences were aligned using MACSE v2.06 [49], and the alignments were manually inspected and trimmed in MEGA 11 [50]. BI trees were reconstructed using MrBayes 3.2.7a [51] implemented in PhyloSuite v1.2.2 [52], and the resulting trees are provided in the Supplementary Materials. Final tree visualization and annotation were performed using iTOL v7.6 [53].

2.6. Differential Expression Analysis

To characterize tissue-specific expression patterns, FPKM values of candidate chemosensory genes were $\log_2$(FPKM + 1) transformed and normalized. The processed data were uploaded to Biomarker's online platform for heatmap visualization using default clustering parameters.

Differential expression analysis was conducted on Biomarker's online platform using the EBSeq package. Genes with $|\log_2(\text{fold change})| > 1$ and $p < 0.05$ were considered significantly differentially expressed. Special emphasis was placed on the comparison between FA and MA to identify female antenna-biased chemosensory candidates.

2.7. qRT-PCR Validation

To support transcriptome-derived expression profiling, qRT-PCR was performed for ten representative odorant receptor genes (*LglyORco*, *LglyOR1a*, *LglyOR11b*, *LglyOR49b*, *LglyOR6c*, *LglyOR6d*, *LglyOR15a*, *LglyOR2*, *LglyOR6b*, and *LglyOR6a*) using an independent batch of male antennae from *L. glycinivorella*. Total RNA from the independent batch of male antennae was treated with DNase I and reverse-transcribed into cDNA. qRT-PCR was performed using PerfectStart® Green qPCR SuperMix (TransGen Biotech, Beijing, China; AQ601-02) according to the manufacturer's instructions, and amplification specificity was confirmed by melting-curve analysis. Relative expression levels were calculated using the $2^{-\Delta\Delta Ct}$ method with Actin as the internal reference gene. Data are presented as mean ± SEM from technical replicates (wells, $n = 9$). Primer sequences are provided in Table S2.

https://doi.org/10.3390/biology15060505

3. Results

3.1. Transcriptome Sequencing and Assembly Quality Assessment

High-throughput transcriptome sequencing was performed on eight tissue samples of *L. glycinivorella*, including female antennae (FA), male antennae (MA), female heads without antennae (FH), male heads without antennae (MH), pooled thoraces (T), pooled legs (L), female abdomens (FAb), and male abdomens (MAb). In total, 137.99 Gb of clean data were generated, with each sample yielding at least 6.04 Gb. The Q30 percentage exceeded 97.78% across all samples, indicating high sequencing quality.

Clean reads were mapped to the *L. glycinivorella* reference genome, with mapping rates ranging from 66.83% to 80.41%. Based on these alignments, alternative splicing events were predicted, gene models were refined, and novel genes were identified. In total, 5593 novel genes were discovered, of which 1402 received functional annotations (Table 1).

Table 1. Summary statistics of sequencing data.

Samples	Clean Reads	Clean Bases	GC Content	≥Q30 (%)
FA	21,027,433	6,296,277,380	0.425733333	0.9796
FAB	20,524,178.67	6,139,020,059	0.439433333	0.984133333
H	21,191,536	6,343,141,690	0.443166667	0.984033333
Leg	21,073,984.67	6,305,114,227	0.419933333	0.9828
MA	21,036,382	6,293,557,836	0.4264	0.9827
MAb	20,726,886.67	6,192,582,732	0.457666667	0.984266667
MH	21,337,162.67	6,384,668,444	0.448233333	0.983366667
T	20,846,713.67	6,237,254,883	0.416766667	0.983566667

(1) Samples: Sample analysis ID; (2) Clean reads: Total number of paired-end reads in clean data; (3) Clean bases: Total number of bases in clean data; (4) GC content: GC percentage in clean data, i.e., the proportion of guanine (G) and cytosine (C) bases among all bases in the clean data; (5) Q30%: Percentage of bases in clean data with a quality score greater than or equal to 30.

3.2. Identification of Chemosensory-Related Genes

Based on transcriptome annotation and homology searches, we identified candidate members of the six major chemosensory gene families, including odorant receptors (ORs), gustatory receptors (GRs), ionotropic receptors (IRs), odorant-binding proteins (OBPs), chemosensory proteins (CSPs), and sensory neuron membrane proteins (SNMPs). Among the soluble olfactory proteins, we identified 52 OBPs and 18 CSPs (Supplementary Tables S3 and S4). All candidate sequences were further validated by ORF prediction. Among them, 44 OBPs and 16 CSPs contained full-length open reading frames. Sequence analysis revealed that most canonical OBPs retain the conserved six-cysteine motif (C-pattern), while CSPs possess four conserved cysteine residues, both of which are important for tertiary structure stability and ligand binding. Both OBP and CSP members were predicted to contain an N-terminal signal peptide and to lack transmembrane domains, consistent with their roles as secreted carrier proteins that bind and transport hydrophobic volatile molecules through the aqueous sensillar lymph.

In addition to soluble proteins, we identified four families of membrane-bound chemosensory genes (Supplementary Tables S5–S8): 76 ORs, of which 54 were full-length; 15 GRs, all full-length; 18 IRs, with 7 full-length; and 4 SNMPs, all full-length. ORs and GRs typically exhibited 6–7 predicted transmembrane domains, consistent with seven-transmembrane chemosensory receptors. SNMPs featured two transmembrane domains and a large extracellular loop, which are hallmark features of CD36-family proteins implicated in olfactory signaling. IRs retained conserved domains and key residues homologous to ionotropic glutamate receptors (iGluRs), reflecting structural conservation. Overall, the structural features of *L. glycinivorella* chemosensory genes closely resemble those reported in other lepidopterans, supporting the reliability of our annotations.

https://doi.org/10.3390/biology15060505

3.3. Phylogenetic Analysis of Chemosensory Genes

To infer evolutionary relationships, maximum-likelihood phylogenetic trees were constructed for all six chemosensory gene families (OR, GR, IR, OBP, CSP, and SNMP) using homologs from representative Lepidoptera. In general, *L. glycinivorella* candidates clustered within well-supported, conserved clades of lepidopteran orthologs, providing phylogenetic support for functional inference. Bayesian inference (BI) phylogenetic analyses based on CDS datasets yielded topologies generally consistent with the ML trees, further supporting the robustness of the inferred phylogenetic relationships (Supplementary Figures S1–S6).

OR Family (Figure 1): The co-receptor *LglyOrco* formed a distinct, strongly supported clade with Orco orthologs from other lepidopterans, consistent with its conserved role. Excluding Orco, 13 LglyORs grouped within the lepidopteran pheromone receptor (PR) clade, suggesting potential involvement in sex pheromone detection. The remaining LglyORs were distributed across multiple subfamilies: some showed putative one-to-one orthology with reported host-volatile receptors, whereas others formed species-specific clusters, indicating possible functional diversification in host-odor and environmental-cue recognition. The BI tree showed a generally consistent topology with the ML tree, further supporting the phylogenetic placement of LglyOrco and the candidate PR lineage (Supplementary Figure S1).

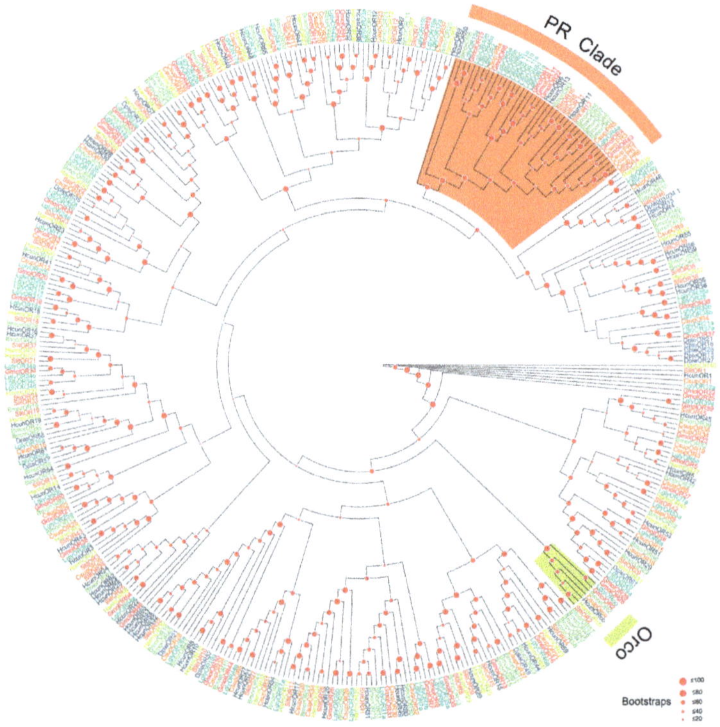

Figure 1. Homology analysis of ORs from *L. glycinivorella* and representative Lepidoptera. The sequences used to construct the phylogenetic tree were derived from the following 7 species: *Grapholita molesta* (Busck, 1916) (Gmol) [54,55]; *Chilo suppressalis* (Csup) [56]; *Helicoverpa armigera* (Harm) [55]; *Bombyx mori* (Linnaeus, 1758) (Bmor); *Danaus plexippus* (Linnaeus, 1758) (Dple); *Hyphantria cunea* (Drury, 1773) (Hcun) [14]; *Spodoptera litura* (Fabricius, 1775) [57]. Different font colors indicate different species, and the shaded colored regions indicate the major clades labeled in the figure.

https://doi.org/10.3390/biology15060505

GR Family (Figure 2): The phylogenetic tree clarified the correspondence between *L. glycinivorella* GR members and major functional subfamilies of lepidopteran GRs. Within the putative CO_2 receptor clade (GR1/GR2/GR3 lineage), *LglyGR2* clustered with orthologs from other lepidopterans, whereas no GR1- or GR3-like homologs were recovered from our transcriptome dataset. In the sugar receptor-related clade, *LglyGR5* and *LglyGR6* were identified. In addition, no *L. glycinivorella* candidates were assigned to the bitter taste receptor-related clade in the phylogenetic analysis. Bayesian inference analysis recovered a topology largely consistent with the ML tree, supporting the inferred placement of the CO_2 receptor and sugar receptor-related GR lineages (Supplementary Figure S2).

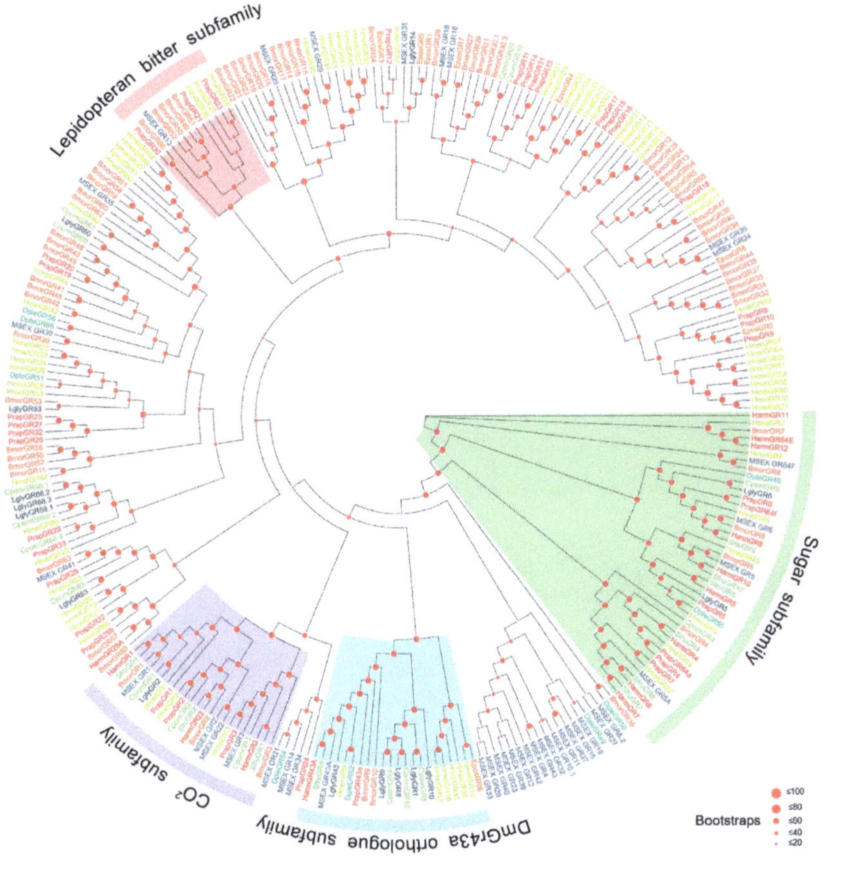

Figure 2. Homology analysis of GRs from *L. glycinivorella* and representative Lepidoptera. The sequences used to construct the phylogenetic tree were derived from the following 9 species: *Cydia pomonella* (Linnaeus, 1758) (Cpom) [58]; *Pieris rapae* (Linnaeus, 1758) (Prap) [59]; *Manduca sexta* (Linnaeus, 1763) (Msex) [60]; *Bombyx mori* (Bmor) [61]; *Heliconius melpomene* (Linnaeus, 1758) (Hmel) [62]; *Helicoverpa armigera* (Harm) [63]; *Spodoptera frugiperda* (J. E. Smith, 1797) (Sfru) [64]; *Danaus plexippus* (Dple) [65]; *Epiphyas postvittana* (Walker, 1863) (Epos) [66]. Different font colors indicate different species, and the shaded colored regions indicate the major clades labeled in the figure.

IR family (Figure 3): Among the 18 identified LglyIR genes, phylogenetic analysis classified them into three subfamilies. The Antennal IRs (A-IRs) include two conserved co-receptors, *LglyIR8a* and *LglyIR25a*, as well as a group of tuning receptors—*LglyIR21a*, *LglyIR40a*, *LglyIR75q.1*, and *LglyIR87a*. Additionally, a *L. glycinivorella*-specific clade was

 https://doi.org/10.3390/biology15060505

identified, which clusters with the co-receptors IR25a and IR8a and comprises four paralogs: *LglyIR25b*, *LglyIR25c*, *LglyIR25d*, and *LglyIR25e*. Within the Lepidoptera-Specific IRs (LS-IRs), three members were identified: *LglyIR1.1*, *LglyIR1.2*, and *LglyIR2*. In the Divergent IRs (D-IRs), only four copies from the IR7d family were detected (*LglyIR7d.1.1*, *LglyIR7d.1.2*, *LglyIR7d.2.1*, and *LglyIR7d.4*). No *L. glycinivorella* homologs were found in the IR100a, IR100b, IR143, or IR85a clades. The BI tree was generally congruent with the ML topology and supported the major IR clades identified in this study (Supplementary Figure S3).

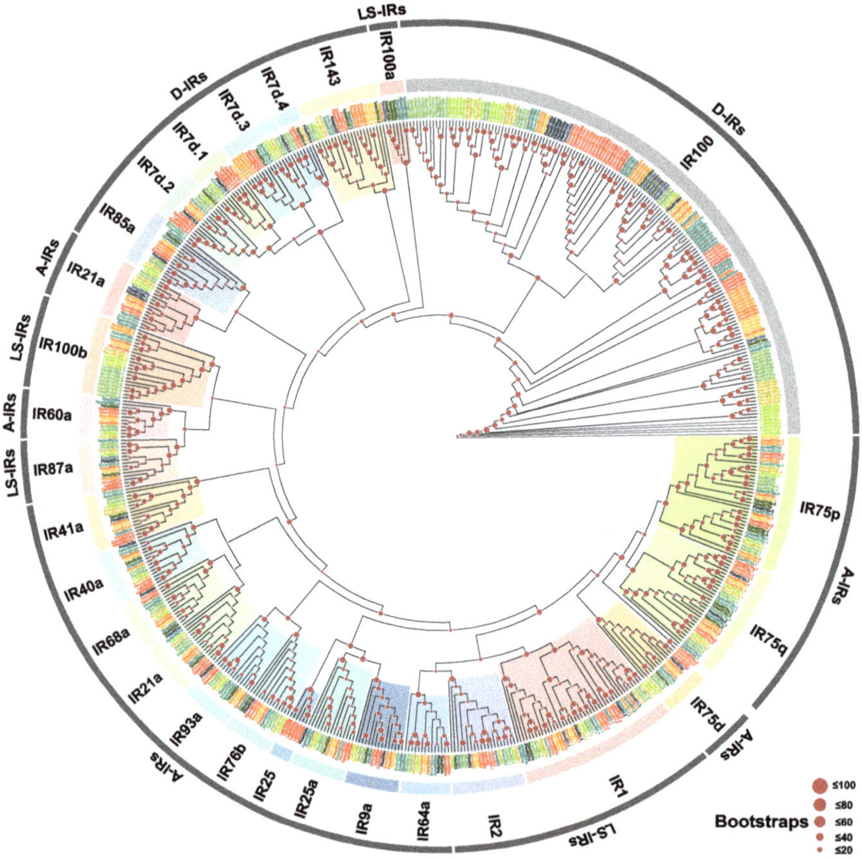

Figure 3. Homology analysis of IRs from *L. glycinivorella* and representative Lepidoptera. The sequences used to construct the phylogenetic tree were derived from the following 14 species: *Cydia pomonella* (Cpom); *Adoxophyes honmai* (Yasuda, 1998) (Ahon); *Papilio bianor* (Cramer, 1777) (Pbia); *Chilo suppressalis* (Csup); *Ostrinia furnacalis* (Guenée, 1854) (Ofur); *Galleria mellonella* (Linnaeus, 1758) (Gmel); *Hyphantria cunea* (Hcun); *Lymantria dispar* (Linnaeus, 1758) (Ldis); *Agrotis ipsilon* (Hufnagel, 1766) (Aips); *Spodoptera exigua* (Hübner, 1808) (Sexi); *Antheraea yamamai* (Guérin-Méneville, 1861) (Ayan); *Bombyx mandarina* (Moore, 1872) (Bman); *Eumeta japonica* (Heylaerts, 1884) (Ejap); *Tuta absoluta* (Meyrick, 1917) (Tabs) [18]. Different font colors indicate different species, and the shaded colored regions indicate the major clades labeled in the figure.

OBP Family (Figure 4): Phylogenetic analysis showed that *L. glycinivorella* OBPs are distributed across the major lepidopteran OBP clades. Notably, several LglyOBPs formed sister pairs or small species-specific clades (e.g., *LglyOBP12a/12b* and *LglyOBP18a/18b*), consistent with recent lineage-specific duplications followed by sequence divergence. Bayesian inference analysis yielded a generally consistent topology with the ML tree, further sup-

 https://doi.org/10.3390/biology15060505

porting the observed OBP clustering pattern and the inferred lineage-specific duplications (Supplementary Figure S4).

Figure 4. Homology analysis of OBPs from *L. glycinivorella* and other representative insect species. The sequences used to construct the phylogenetic tree were derived from the following 10 species: *Plodia interpunctella* (Hübner, 1813) (Pint); *Ostrinia furnacalis* (Ofur); *Conogethes punctiferalis* (Guenée, 1854) (Cpun); *Chilo suppressalis* (Csup); *Bombyx mori* (Bmor) [67]; *Grapholita molesta* (Gmol); *Spodoptera litura* (Slit); *Helicoverpa armigera* (Harm) [54]; *Cydia pomonella* (Cpom) [68]; *Mythimna separata* (Walker, 1865) (Msep) [69]. Different font colors indicate different species.

CSP family (Figure 5): *L. glycinivorella* CSPs co-clustered with CSPs from representative lepidopteran species across multiple clades, indicating an overall conserved evolutionary pattern. A few members formed closely related pairs or species-specific subclades (e.g., *LglyCSP7a* and *LglyCSP7b*), suggesting possible local gene duplication or expansion. Such lineage-specific changes may contribute to the functional diversification of CSPs beyond olfaction across multiple tissues. The BI tree showed an overall topology consistent with the ML analysis, supporting both the conserved CSP clades and the inferred species-specific expansions (Supplementary Figure S5).

 https://doi.org/10.3390/biology15060505

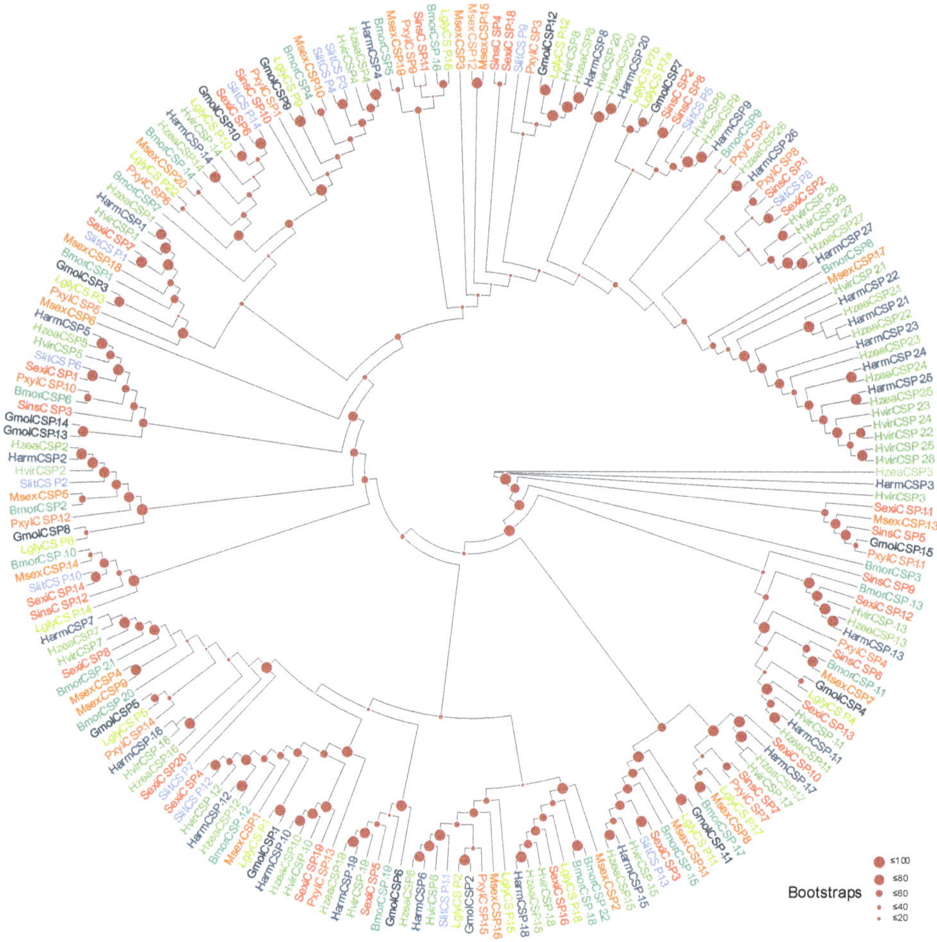

Figure 5. Homology analysis of CSPs from *L. glycinivorella* and other representative insect species. The sequences used to construct the phylogenetic tree were derived from the following 10 species: *Grapholita molesta* (Gmol); *Spodoptera litura* (Slit) [54]; *Streltzoviella insularis* (Staudinger, 1892) (Sins); *Spodoptera exigua* (Sexi) [70]; *Bombyx mori* (Bmor); *Helicoverpa armigera* (Harm); *Heliothis virescens* (Fabricius, 1777) (Hvir); *Plutella xylostella* (Linnaeus, 1767) (Pxyl); *Helicoverpa zea* (Boddie, 1850) (Hzea); *Manduca sexta* (Msex) [71]. Different font colors indicate different species.

SNMP Family (Figure 6): SNMP members were clearly divided into two subfamilies, SNMP1 and SNMP2. *LglySNMP1* clustered with SNMP1 homologs from other lepidopterans in a single well-supported clade, consistent with the reported association of SNMP1 with sex pheromone detection. In contrast, *LglySNMP2a*, *LglySNMP2b*, *LglySNMP3c* grouped within the SNMP2 clade, reflecting divergence from SNMP1. Given that SNMP1 has been shown to play an essential role in sex pheromone perception in several lepidopteran species, *LglySNMP1* represents a high-priority candidate for subsequent functional validation. Bayesian inference analysis also supported the separation of SNMP1 and SNMP2 and the phylogenetic placement of LglySNMP1 within the SNMP1 clade (Supplementary Figure S6).

https://doi.org/10.3390/biology15060505

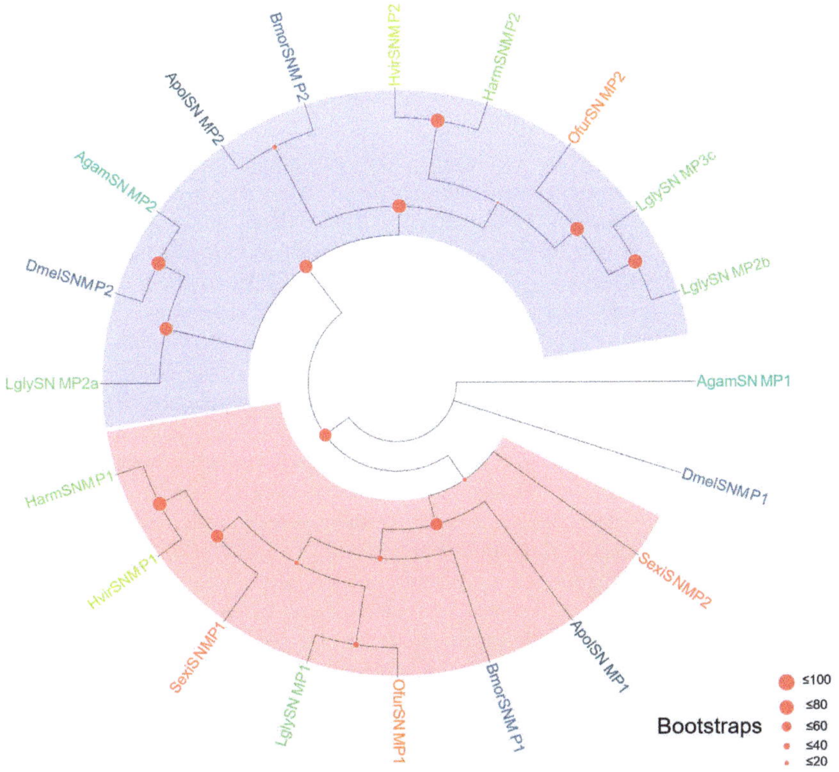

Figure 6. Homology analysis of sensory neuron membrane proteins (SNMPs) from *L. glycinivorella* and other representative insect species. The sequences used to construct the phylogenetic tree were derived from the following 8 species: *Anopheles gambiae* (Giles, 1900) (Agam); *Bombyx mori* (Bmor); *Drosophila melanogaster* (Meigen, 1830) (Dmel); *Helicoverpa armigera* (Harm); *Heliothis virescens* (Hvir); *Ostrinia furnacalis* (Ofur); *Spodoptera exigua* (Sexi); *Antheraea polyphemus* (Cramer, 1775) (Apol) [72]. Different font colors indicate different species, and the shaded colored regions indicate the major clades labeled in the figure.

In summary, phylogenetic analyses indicate that the chemosensory-related genes of *L. glycinivorella* exhibit typical lepidopteran evolutionary characteristics and show evidence of potential gene duplication and lineage-specific divergence, providing a foundation for subsequent screening of key functional candidates based on expression profiles, electrophysiological assays, or genetic manipulation. The overall congruence between the ML and BI trees further supports the robustness of these phylogenetic inferences.

3.4. Tissue-Specific Expression Analysis of Chemosensory Genes

To characterize the expression distribution of chemosensory-related genes across tissues and sex-biased expression in the antennae of *L. glycinivorella*, we generated clustered heatmaps based on FPKM values for the six gene families (OR, GR, IR, OBP, CSP, and SNMP) and performed differential expression analysis between female and male antennae; significantly differentially expressed genes were visualized in volcano plots.

Overall, the heatmap results showed that, compared with heads (antennae removed), thoraces, legs, and abdomens, genes from the OR, IR, OBP, and SNMP families were generally more highly expressed in antennae, exhibiting a clear antennal-enriched pattern. In contrast,

https://doi.org/10.3390/biology15060505

CSP and GR family genes showed broader expression overall, although some members still exhibited relatively higher expression in antennae or displayed sex-biased expression.

At the family level, the OR family was strongly enriched in antennae, and several ORs showed relatively higher expression in female antennae (e.g., *LglyOR13* and *LglyOR15*). The GR family exhibited a relatively dispersed expression pattern but still included sex-differentially expressed genes. The IR family contained only a small number of differentially expressed genes, and no down-regulated genes were detected. Among soluble carrier proteins, the OBP family was primarily antennal-enriched and included sex-differentially expressed genes (e.g., *LglyOBP2* and *LglyOBP8*), whereas the CSP family was broadly expressed, with only a few genes showing sex-biased expression. Finally, the SNMP family showed overall high expression in antennae and a detectable degree of sex-biased expression.

Multiple OR genes in *L. glycinivorella* were significantly enriched in male antennae, whereas several genes, including *LglyOR13*, *LglyOR15*, *LglyOR32a*, and *LglyOR35a*, exhibited tissue-specific high expression in female antennae, heads, or abdomens (Figure 7A). The differential-expression volcano plot (Figure 7B) further indicated sex-biased expression between female and male antennae: 12 OR genes were significantly higher in male antennae (e.g., *LglyOR11e*, *LglyOR2*, *LglyOR6c*, and *LglyOR4a*), whereas five genes (e.g., *LglyOR10*, *LglyOR32a*, and *LglyOR35a*) showed significantly lower expression in female antennae; most remaining ORs did not exhibit significant changes.

To strengthen the robustness of transcriptome-based expression profiling, we performed qRT-PCR for ten representative OR genes using an independent batch of male antennae (Figure 7C). The qRT-PCR results revealed a clear expression hierarchy among these candidates in male antennae, providing independent experimental support for the transcriptome-derived OR expression profile. ORs were prioritized for validation because OR-mediated olfaction constitutes a core molecular basis of odor detection in insects and is directly relevant to antennal odorant recognition.

Most LglyGR genes showed higher expression in the antennae, particularly in male antennae, such as *LglyGR5*, *LglyGR6*, *LglyGR2*, and *LglyGR43*. In contrast, *LglyGR68.1*, *LglyGR63*, *LglyGR60*, *LglyGR53*, *LglyGR8*, *LglyGR14*, *LglyGR10*, *LglyGR68.2*, *LglyGR68.3*, and *LglyGR9* exhibit higher expression in thorax, legs, or female tissues (Figure 8A). Some of these genes are expressed more highly in female antennae or abdomen, indicated in blue, suggesting they may participate in gender-specific or non-olfactory-related sensory functions. Further analysis using a volcano plot examined the differential expression of *L. glycinivorella* GR genes between male and female antennae (Figure 8B). It was found that *LglyGR53* and *LglyGR68.1* were significantly upregulated in male antennae, while *LglyGR14* and *LglyGR43* were significantly downregulated in female antennae. The remaining genes did not show significant differential expression.

Among the antennal IRs (A-IRs), the co-receptor *LglyIR8a* was highly expressed in both male and female antennae and showed detectable expression in heads, abdomens, and legs, whereas *LglyIR25a* exhibited extremely low expression (Figure 9A,B). Its paralogs *LglyIR25b*–e displayed divergent expression patterns: *LglyIR25c* was highly expressed in male antennae; *LglyIR25b* was enriched in thoraces and legs; LglyIR25d was primarily expressed in male abdomens; and *LglyIR25e* showed the highest expression in legs. Among the tuning receptors, *LglyIR75q*.1 showed the highest expression in male antennae, and LglyIR40a was also enriched in male antennae, whereas *LglyIR21a* and *LglyIR87a* were expressed at low levels in antennae; Within the lepidopteran-specific IRs (LS-IRs), *LglyIR2* showed the highest expression in male antennae, followed by *LglyIR1.2*, while *LglyIR1*.1 was expressed at a low level in male antennae; all three showed very low or undetectable expression in other tissues. In the divergent IRs (D-IRs), *LglyIR7d.2.1* displayed the highest expression in male antennae, and *LglyIR7d.1.1*, *LglyIR7d.1.2*, and

 https://doi.org/10.3390/biology15060505

LglyIR7d.4 also showed male antennae as their primary expression site, with minimal or no expression in other tissues. Consistent with the phylogenetic analysis, no IR100a-, IR100b-, IR143-, or IR85a-like homologs were recovered from our transcriptome dataset, and no corresponding expression signals were observed in the heatmap. Differential expression analysis between female and male antennae further indicated that *LglyIR25b* and *LglyIR25d* were significantly upregulated in male antennae, with *LglyIR25b* showing the strongest statistical support; no other IR genes showed significant sex-biased expression in antennae.

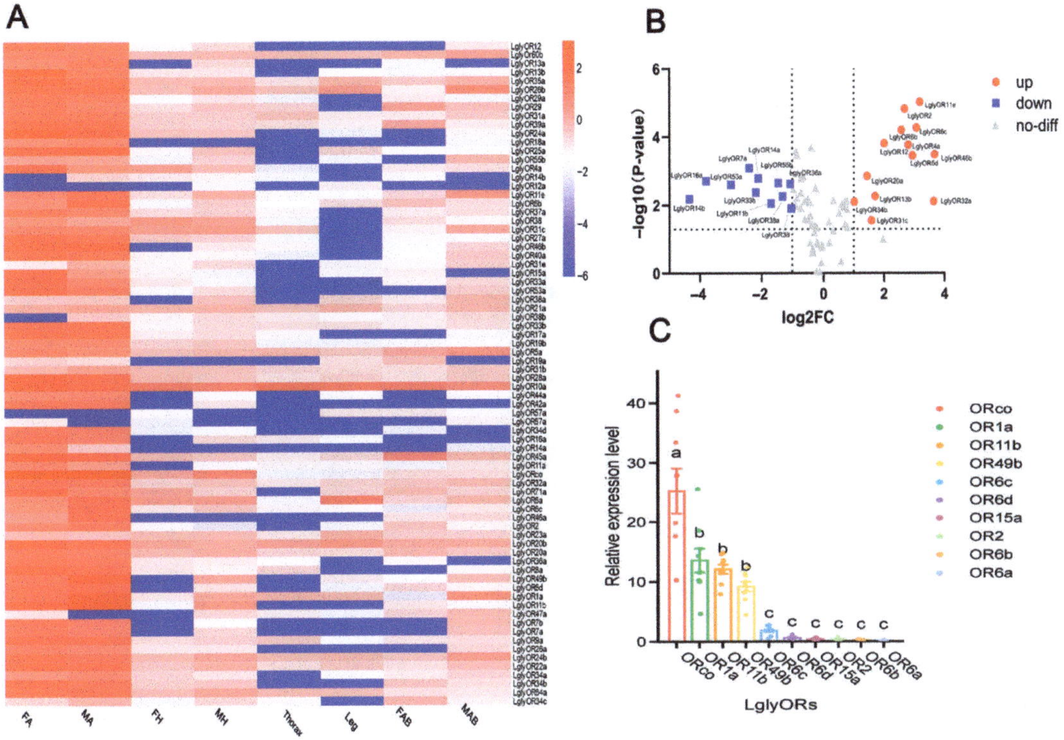

Figure 7. Expression profiles and sex-biased differential expression of OR genes in *L. glycinivorella*, and qRT-PCR assessment of representative candidates. (**A**): Heatmap showing the expression of OR genes in different tissues of *L. glycinivorella*. The color scale ranges from blue (low expression) to red (high expression), representing the normalized expression levels of log10(FPKM). (**B**): Volcano plot of differentially expressed OR genes in *L. glycinivorella*. The horizontal axis represents the log2 fold change in the comparison between female antennae (FA) and male antennae (MA), while the vertical axis shows the −log10(p-value). The vertical dashed lines indicate the thresholds of log2FC = −1 and 1, and the horizontal dashed line indicates the significance threshold of p = 0.05. Red dots indicate significantly differentially expressed genes ($|\log 2FC| > 1$, $p < 0.05$), and gray triangles represent genes without significant differential expression. (**C**): qRT-PCR assessment of ten representative OR genes in male antennae. Relative transcript levels of *LglyORco*, *LglyOR1a*, *LglyOR11b*, *LglyOR49b*, *LglyOR6c*, *LglyOR6d*, *LglyOR15a*, *LglyOR2*, *LglyOR6b*, and *LglyOR6a* were quantified using the $2^{-\Delta\Delta Ct}$ method with actin as the reference gene. Bars represent mean ± SEM, and dots indicate individual qPCR wells (technical replicates, n = 9). Different letters denote significant differences among genes (one-way ANOVA with Tukey's multiple comparisons, $p < 0.05$; based on technical replicates).

https://doi.org/10.3390/biology15060505

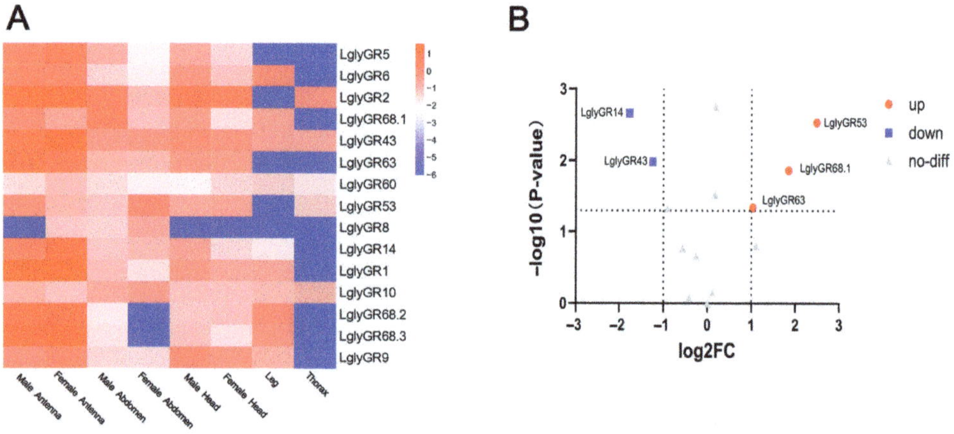

Figure 8. Tissue expression profiles and differential expression of gustatory receptor (GR) genes in *L. glycinivorella*. (**A**): Heatmap showing the expression of GR genes in different tissues of *L. glycinivorella*. The color scale ranges from blue (low expression) to red (high expression), representing the normalized expression levels of log10(FPKM). (**B**): Volcano plot of differentially expressed GR genes in *L. glycinivorella*. The horizontal axis represents the log2 fold change in the comparison between female antennae (FA) and male antennae (MA), while the vertical axis shows the −log10(*p*-value). The vertical dashed lines indicate the thresholds of log2FC = −1 and 1, and the horizontal dashed line indicates the significance threshold of $p = 0.05$. Red dots indicate significantly differentially expressed genes (|log2FC| > 1, $p < 0.05$), and gray triangles represent genes without significant differential expression.

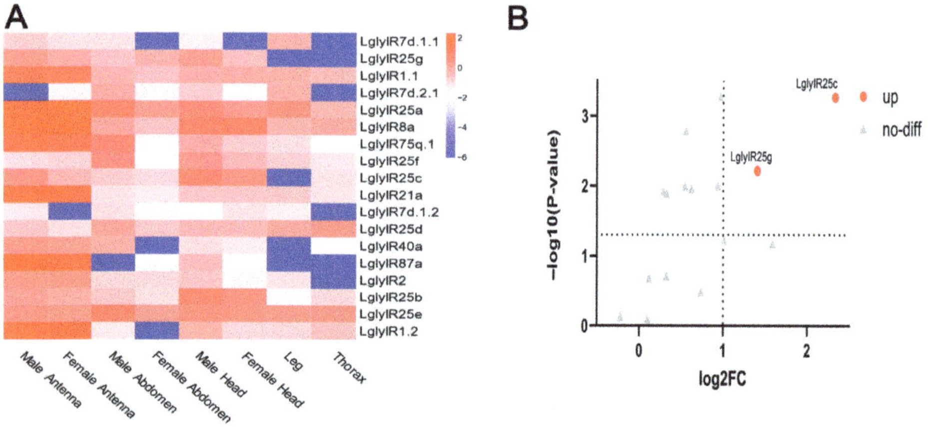

Figure 9. Tissue expression profiles and differential expression of ionotropic receptor (IR) genes in *L. glycinivorella*. (**A**): Heatmap showing the expression of IR genes in different tissues of *L. glycinivorella*. The color scale ranges from blue (low expression) to red (high expression), representing the normalized expression levels of $\log_{10}$(FPKM). (**B**): Volcano plot of differentially expressed IR genes in *L. glycinivorella*. The horizontal axis represents the log2 fold change in the comparison between female antennae (FA) and male antennae (MA), while the vertical axis shows the −log10(*p*-value). The vertical dashed lines indicate the thresholds of log2FC = 1, and the horizontal dashed line indicates the significance threshold of $p = 0.05$. Red dots indicate significantly differentially expressed genes (|log2FC| > 1, $p < 0.05$), and gray triangles represent genes without significant differential expression.

https://doi.org/10.3390/biology15060505

The heatmap showed that most LglyOBP genes were highly expressed in antennae, with pronounced enrichment in both male and female antennae (Figure 10A). Notably, several OBPs (e.g., *LglyOBP33*, *LglyOBP4*, and *LglyOBP23*) exhibited higher expression in male antennae. Volcano-plot analysis further indicated that multiple OBP genes were significantly differentially expressed between antennae and non-antennal tissues, with *LglyOBP33*, *LglyOBP4*, and *LglyOBP23* significantly upregulated and *LglyOBP19* and *LglyOBP12a* significantly downregulated (Figure 10B).

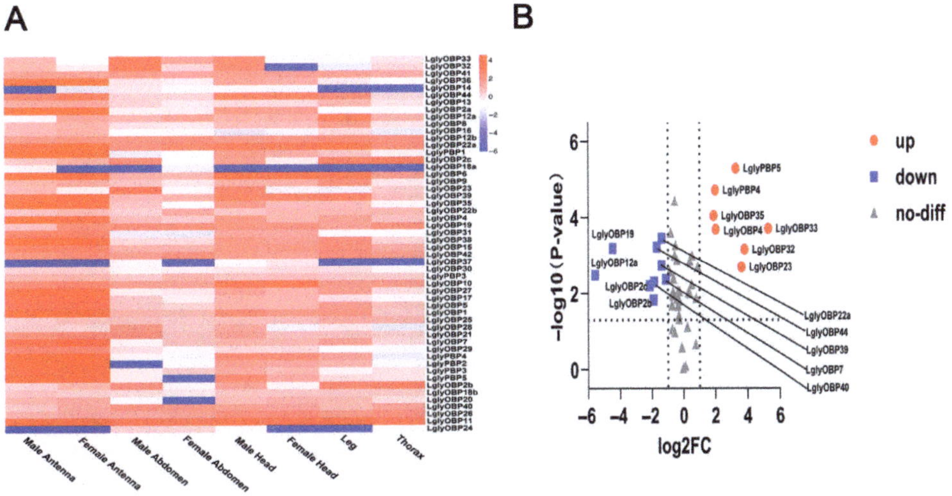

Figure 10. Tissue expression profiles and differential expression of odorant-binding protein (OBP) genes in *L. glycinivorella*. (**A**): Heatmap showing the expression of OBP genes in different tissues of *L. glycinivorella*. The color scale ranges from blue (low expression) to red (high expression), representing the normalized expression levels of $\log_{10}$(FPKM). (**B**): Volcano plot of differentially expressed OBP genes in *L. glycinivorella*. The horizontal axis represents the log2 fold change in the comparison between female antennae (FA) and male antennae (MA), while the vertical axis shows the −log10(*p*-value). The vertical dashed lines indicate the thresholds of log2FC = −1 and 1, and the horizontal dashed line indicates the significance threshold of $p = 0.05$. Red dots indicate significantly differentially expressed genes ($|\text{log2FC}| > 1$, $p < 0.05$), and gray triangles represent genes without significant differential expression.

The heatmap showed that most LglyCSP genes were highly expressed in the antennae of *L. glycinivorella*, with clear sex-biased patterns. *LglyCSP1*, *LglyCSP2*, *LglyCSP8*, *LglyCSP11*, and *LglyCSP14* exhibited higher expression in male antennae, whereas *LglyCSP15* and *LglyCSP17* showed higher expression in female antennae (Figure 11A). In addition, some genes (e.g., *LglyCSP2* and *LglyCSP3*) displayed relatively high expression across multiple tissues. Volcano-plot analysis further indicated that *LglyCSP1* and *LglyCSP22* were significantly upregulated in male antennae, whereas *LglyCSP17* was significantly upregulated in female antennae (Figure 11B).

LglySNMP1 exhibited significantly high expression in both male and female antennae, while its expression was markedly lower in other tissues such as the abdomen, head, and thorax. *LglySNMP2a* and *LglySNMP2b* also showed relatively high expression in antennae and head, but their expression was broader compared to *LglySNMP1*, extending to legs and thorax (Figure 12A). The expression pattern of *LglySNMP3c* resembled that of *LglySNMP1*, though it was nearly absent in the abdomen and thorax. Volcano plot analysis of specific comparison groups (e.g., antennae vs. other tissues) further revealed that only *LglySNMP1* was significantly upregulated in this contrast, whereas the other three genes did not show significant differential expression (Figure 12B).

 https://doi.org/10.3390/biology15060505

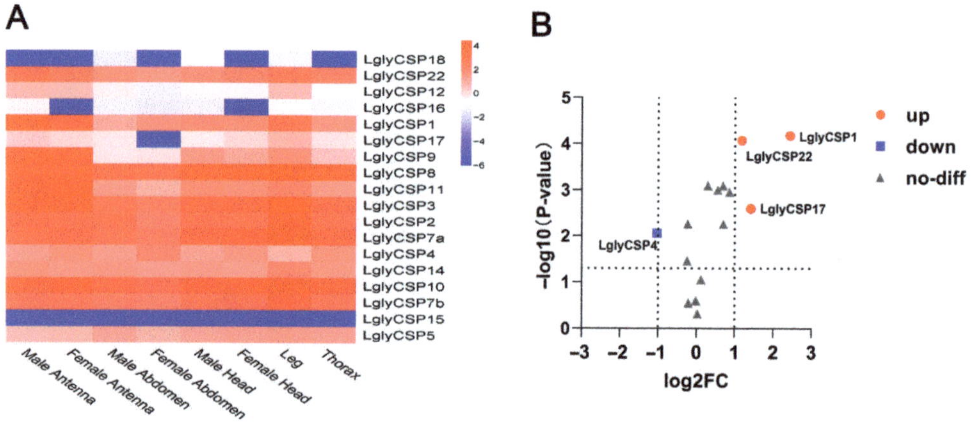

Figure 11. Tissue expression profiles and differential expression of chemosensory protein (CSP) genes in *L. glycinivorella*. (**A**): Heatmap showing the expression of CSP genes in different tissues of *L. glycinivorella*. The color scale ranges from blue (low expression) to red (high expression), representing the normalized expression levels of log10(FPKM). (**B**): Volcano plot of differentially expressed CSP genes in *L. glycinivorella*. The horizontal axis represents the log2 fold change in the comparison between female antennae (FA) and male antennae (MA), while the vertical axis shows the −log10(*p*-value). The vertical dashed lines indicate the thresholds of log2FC = −1 and 1, and the horizontal dashed line indicates the significance threshold of $p = 0.05$. Red dots indicate significantly differentially expressed genes (∣log2FC∣ > 1, $p < 0.05$), and gray triangles represent genes without significant differential expression.

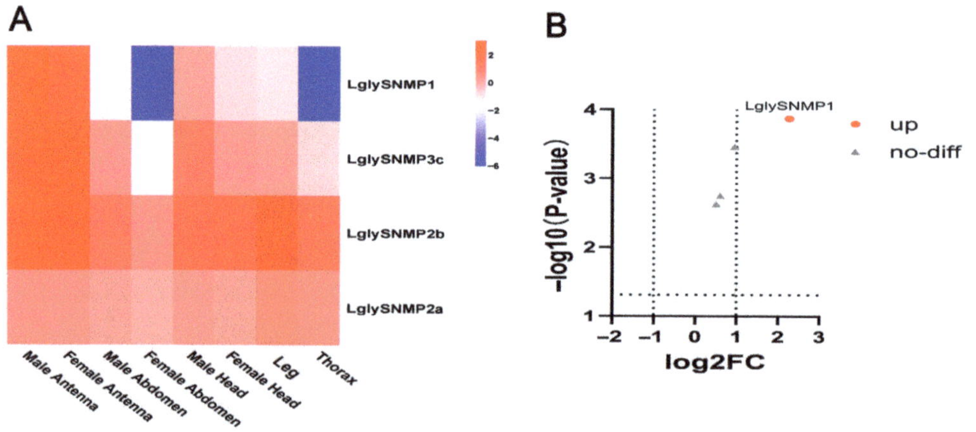

Figure 12. Tissue expression profiles and differential expression of sensory neuron membrane protein (SNMP) genes in *L. glycinivorella*. (**A**): Heatmap showing the expression of SNMP genes in different tissues of *L. glycinivorella*. The color scale ranges from blue (low expression) to red (high expression), representing the normalized expression levels of log10(FPKM). (**B**): Volcano plot of differentially expressed SNMP genes in *L. glycinivorella*. The horizontal axis represents the log2 fold change in the comparison between female antennae (FA) and male antennae (MA), while the vertical axis shows the −log10(*p*-value). The vertical dashed lines indicate the thresholds of log2FC = −1 and 1, and the horizontal dashed line indicates the significance threshold of $p = 0.05$.Red dots indicate significantly differentially expressed genes (∣log2FC∣ > 1, $p < 0.05$), and gray triangles represent genes without significant differential expression.

https://doi.org/10.3390/biology15060505

4. Discussion

A thorough dissection of the chemosensory system in *L. glycinivorella* provides molecular targets and a theoretical basis for behavior-based green pest management strategies within integrated pest management (IPM), such as pheromone trapping and the screening of host-volatile attractants or repellents. Chemosensory molecules—including ORs, GRs, IRs, and peripheral auxiliary proteins such as OBPs—represent key entry points for "reverse chemical ecology" approaches aimed at discovering semiochemicals and optimizing behavioral control tactics [73]. However, the molecular mechanisms underlying chemical perception in *L. glycinivorella* remain poorly understood. Here, we systematically identified chemosensory-related genes through multi-tissue transcriptome sequencing and bioinformatic analyses and characterized their sequence features, phylogenetic relationships, and tissue-specific expression patterns, thereby providing a resource for subsequent functional validation and ligand screening of candidate genes [74].

ORs are core receptors in insect olfaction and typically form heteromeric complexes with the conserved co-receptor ORco to mediate signal transduction. In this study, we identified LglyORco, providing a basis for subsequent heterologous expression and ligand screening. In the silkworm (*Bombyx mori*), loss of ORco significantly reduces olfactory sensitivity and impairs related behaviors, underscoring the functional conservation and essentiality of Orco [75]. In Lepidoptera, pheromone receptors (PRs) represent a specialized clade within the OR family and mediate mate recognition. For instance, *BmOR1* in *B. mori* has been functionally validated to detect a sex pheromone component [76], and PR lineages have undergone species-specific expansion and functional diversification across moth species [77]. Here, we identified 13 candidate PRs in *L. glycinivorella*, indicating a potentially diverse receptor repertoire for sex pheromone detection; however, functional assignment will require validation using sex- and tissue-specific expression patterns and ligand-response profiles from heterologous expression assays. A limitation of this study is the lack of paired qRT-PCR validation for FA–MA DEGs and the absence of functional assays for candidate ORs. Therefore, the PR candidates and female-biased ORs proposed here should be regarded as hypothesis-driven priorities for subsequent experimental validation. In contrast, several ORs in our dataset (e.g., *LglyOR13* and *LglyOR15*) were upregulated in female antennae, more plausibly indicating roles in host location or oviposition-site selection. Notably, phylogenetic placement alone does not strictly predict ligand type: for example, *CpomOR3* in codling moth (*Cydia pomonella*), initially annotated as PR-like, was shown to respond to the host volatile pear ester [78]; similarly, *HassOR31* in *Helicoverpa assulta* (Guenée, 1852) is highly expressed in the ovipositor and is implicated in host-volatile detection [79]. Therefore, as working hypotheses, candidate ORs can be prioritized into two functional categories: (1) PR candidates likely associated with sex pheromone communication, and (2) female-biased ORs potentially involved in recognizing host- or oviposition-related chemical cues. This framework provides biologically plausible priorities for future functional validation.

GRs primarily mediate insect perception of contact chemical stimuli—such as sugars and bitter compounds—as well as gaseous cues such as CO_2, providing a molecular basis for feeding selection, host recognition, and oviposition decisions [63]. In this study, we identified 15 candidate LglyGR genes from multi-tissue transcriptomes, including putative GR43a-like ("fructose receptor") members, candidate CO_2 receptor-related genes, and sugar receptor-related members, suggesting that *L. glycinivorella* retains molecular components for detecting nutritional signals and CO_2-related cues [63]. Functional evidence from other insects supports these annotations. In Drosophila, Gr43a has been confirmed as a highly specific fructose receptor and internal nutrient sensor that modulates feeding in response to physiological fructose levels [80]. Similar functions have been reported in Lepidoptera:

https://doi.org/10.3390/biology15060505

Helicoverpa armigera HaGr4 responds specifically to fructose [81], while *HaGr9* (reported as identical to or closely related to *HaGr4*) also responds to D-maltose and D-galactose in addition to fructose [82]. Likewise, *Ostrinia furnacalis* OfurGr43 has been shown to detect both D-fructose and D-sorbitol [61]. In *L. glycinivorella*, *LglyGR1*, *LglyGR8*, *LglyGR9*, *LglyGR10*, and *LglyGR43* were assigned to the GR43a-related clade, suggesting that this pest may exhibit sensitivity to fructose-associated cues.

Notably, we did not recover homologs assigned to the canonical bitter GR lineage from our dataset [83], which may reflect its specialized oligophagous lifestyle. Regarding CO_2 perception, lepidopterans typically retain a conserved CO_2 receptor lineage composed of three subunits (GR1, GR2, and GR3), which are predominantly expressed in labial palp sensilla and mediate behavioral responses to CO_2 [63]. For example, in *H. armigera*, co-expression of *HarmGr1* with HarmGr3 (or *HarmGr2*) in heterologous systems elicits robust electrophysiological responses to bicarbonate/CO_2, and all three receptors are co-expressed in the same labial palp neuron, supporting a "GR complex-mediated CO_2 detection" model [84]. Similarly, in the fall webworm *Hyphantria cunea*, a binary system expressing *HcunGR1* and *HcunGR3* produces concentration-dependent responses to high CO_2, while a ternary combination (*HcunGR1/2/3*) is activated by CO_2-containing solutions [85]. In *L. glycinivorella*, however, we identified only *LglyGR2* and did not recover GR1- or GR3-like homologs. This pattern may reflect one or more of the following: (1) labial palps were not separately sequenced, resulting in failure to recover low-abundance GR1/GR3 transcripts; (2) CO_2 detection may rely on GR2 alone or GR2 in combination with non-canonical subunits; or (3) CO_2-sensing capability may have undergone functional reduction. Accordingly, *LglyGR2* should be interpreted cautiously as a potential component of a CO_2-sensing module. Future work should prioritize verifying its expression enrichment in key organs such as labial palps and conducting functional assays using CO_2 or bicarbonate as primary candidate stimuli [84].

In this study, phylogenetic analysis of the 18 identified LglyIR genes in *L. glycinivorella* assigned them to the three IR subfamilies defined by Yin et al. (2021): antennal IRs (A-IRs), lepidopteran-specific IRs (LS-IRs), and divergent IRs (D-IRs) [18]. Within the A-IRs, *LglyIR8a* and *LglyIR25a* clustered with conserved co-receptors, whereas *LglyIR21a*, *LglyIR40a*, *LglyIR75q.1*, and *LglyIR87a* fell into their respective tuning receptor clades [25]. Notably, we recovered a *L. glycinivorella*-specific clade comprising *LglyIR25b–e* that grouped with IR25a, suggesting lineage-specific duplication of IR25 in this species. Expression analysis further indicated divergence within this clade: *LglyIR25c* was highly expressed in male antennae; *LglyIR25b* was enriched in thoraces and legs; *LglyIR25d* was primarily expressed in male abdomens; and *LglyIR25e* showed the highest expression in legs. Differential expression analysis further showed that *LglyIR25b* and *LglyIR25d* were significantly upregulated in male antennae, suggesting potential roles in male-biased chemosensory processes. Several tuning receptors exhibited high expression in male antennae, including *LglyIR75q.1*, *LglyIR2*, and *LglyIR40a*, suggesting potential involvement in male-biased behaviors such as mate seeking and host-related odor detection. Within the LS-IRs, in addition to LglyIR2, both *LglyIR1.1* and *LglyIR1.2* were predominantly expressed in male antennae and showed minimal or no expression in other tissues, consistent with lepidopteran-specific olfactory tuning receptors. In the D-IRs, only four members of the IR7d family (*LglyIR7d.1.1*, *LglyIR7d.1.2*, *LglyIR7d.2.1*, and *LglyIR7d.4*) were recovered, and all were primarily expressed in male antennae. Typical D-IR lineages such as IR100a, IR100b, IR143, and IR85a were not recovered from our dataset, and no corresponding expression signals were observed in the heatmap. Given that D-IRs in insects such as Drosophila are mainly implicated in gustation and perception of non-volatile compounds [28–30], this pattern may be associated with extreme host specialization in *L. glycinivorella*. Functionally, IRs often operate as heteromeric

 https://doi.org/10.3390/biology15060505

complexes composed of a co-receptor and a tuning receptor to detect acidic volatiles. For example, the Drosophila IR64a/IR8a complex responds to acidic odors [86]; in mosquitoes, IR8a-mediated pathways contribute to the detection of human-derived acidic compounds such as lactic acid [87]; and in *H. armigera*, IR8a is required for acetic acid attraction [88]. In summary, although the IR repertoire of *L. glycinivorella* is smaller than that of many other lepidopterans [18], its members exhibit pronounced tissue specificity and sex-biased expression, particularly the enrichment of multiple tuning receptors in male antennae, while retaining candidates potentially involved in acid and microenvironmental sensing. These features may reflect adaptation to the chemosensory ecology of its soybean host.

We identified 52 LglyOBP genes, including canonical forms and atypical subtypes (Minus-C, Dimer-C). The reduced cysteine motifs in Minus-C OBPs likely alter ligand-binding pocket flexibility, facilitating functional diversification [89,90]. Expression profiling revealed that *LglyOBP2* and *LglyOBP8* are antenna-enriched, with *LglyOBP8* showing significant female bias. This pattern suggests a specific role in detecting host volatiles or oviposition cues, consistent with female-biased OBPs in other lepidopterans [91]. However, as expression patterns are correlative, future ligand-binding assays and behavioral tests are required to validate their specific ligand spectra and rule out functional redundancy [92].

We observed broad tissue distribution for LglyCSP members, supporting their dual roles in both chemosensation and non-olfactory physiological processes (e.g., development, immunity, and xenobiotic response) [93]. While some CSPs in other species mediate olfactory signaling or dietary adaptation [94,95], the ubiquitous expression of LglyCSPs implies a broader functional divergence beyond odorant transport. Additionally, although CSPs have been linked to insecticide resistance in other insects [96], their specific contributions to *L. glycinivorella*'s environmental adaptation and stress response remain to be elucidated through direct functional validation.

SNMPs belong to the CD36-related family of transmembrane proteins and are typically classified into two subfamilies in the insect peripheral olfactory system: SNMP1 and SNMP2 [35]. In this study, we identified one *LglySNMP1* and three *LglySNMP2* genes in *L. glycinivorella*. Phylogenetic analysis showed that *LglySNMP1* clustered with lepidopteran SNMP1 homologs that have been implicated in sex pheromone detection. Consistently, *LglySNMP1* showed higher expression in male antennae, in line with the reported role of SNMP1 in pheromone-sensitive olfactory neurons and pheromone signal transduction [35, 72]. For example, in *Helicoverpa armigera*, CRISPR-mediated knockout of SNMP1 markedly reduced male electrophysiological responses to long-chain sex pheromone components and impaired upwind flight orientation and behavioral responses to calling females, supporting an essential role of SNMP1 in sex pheromone detection [72]. Therefore, *LglySNMP1* represents a high-priority candidate for functional validation of sex pheromone signaling in *L. glycinivorella*.

In contrast, the three *LglySNMP2* genes identified here exhibited broader expression across multiple tissues, suggesting that their functions may extend beyond direct involvement in sex pheromone recognition [72]. In various moth species, SNMP2 is mainly localized in support cells of olfactory sensilla and has been proposed to participate in transport or clearance of lipophilic compounds, including pheromone degradation products, from the sensillar lymph, thereby contributing to microenvironmental homeostasis and signal termination. Experimental evidence from *Heliothis virescens* and *Bombyx mori* supports a role of SNMP2 in uptake of long-chain fatty acids (pheromone catabolites) and sensillar "lymph clearance" processes [97,98]. Accordingly, *LglySNMP2* members are more likely to function in supportive roles within the peripheral olfactory system, such as maintaining sensillar microenvironment stability and mediating clearance mechanisms. Future studies could combine cellular localization (e.g., in situ hybridization), RNAi or

 https://doi.org/10.3390/biology15060505

CRISPR-based knockdown/knockout, and integrated electrophysiological and behavioral assays to further dissect the distinct roles of *LglySNMP1* and *LglySNMP2* [72,97].

5. Conclusions

In summary, this study systematically identified the chemosensory gene repertoire of *L. glycinivorella* based on multi-tissue transcriptome sequencing, including 76 ORs, 15 GRs, 18 IRs, 52 OBPs, 18 CSPs, and 4 SNMPs. We comprehensively characterized their sequence features, phylogenetic relationships, and tissue-specific expression profiles. The results indicate that, while retaining core olfactory and gustatory pathways, this monophagous pest exhibits marked functional specialization and contraction of several chemosensory gene families relative to polyphagous lepidopterans, and multiple canonical lineages—such as bitter GRs and several divergent IR lineages (e.g., IR100/IR85a)—were not recovered from our dataset. This streamlining may reflect adaptation to the chemical ecology of its sole host plant, soybean. Concurrently, key receptors and auxiliary proteins displayed pronounced differentiation across sexes and tissues: male antennae were enriched for numerous ORs, IRs, and *LglySNMP1*, suggesting important roles in sex pheromone communication and host-volatile perception; in contrast, female-biased or abdomen-enriched expression of genes such as *LglyOBP2*, *LglyOBP8*, *LglyOR13* and *LglyOR15* points to potential involvement in host location and oviposition decisions. This study further highlights functionally informative candidates: male-biased *LglySNMP1* and candidate PRs represent priorities for validation in pheromone- or host volatile-based behavioral manipulation strategies, whereas female- or abdomen-enriched genes offer entry points for disrupting oviposition behavior. In addition, the presence of a GR43a-related clade supports the capacity for detecting nutritional cues such as fructose, whereas the incomplete CO_2 receptor module in our dataset (only *LglyGR2* recovered) suggests potential divergence in CO_2-related sensing. Overall, delineating the composition and expression landscape of chemosensory genes in *L. glycinivorella* provides genetic resources and a conceptual basis for green pest control strategies grounded in "reverse chemical ecology." Future efforts should prioritize ligand deorphanization, neural localization, and behavioral validation to translate these molecular candidates into field applications.

Supplementary Materials: The following supporting information can be downloaded at: https://www.mdpi.com/article/10.3390/biology15060505/s1, Table S1: BioSample and SRA accession numbers for transcriptome samples of *L. glycinivorella*. Table S2: Primers used for qRT-PCR analysis of selected odorant receptor genes in *L. glycinivorella*. Table S3: List of candidate CSP genes in *L. glycinivorella*. Table S4: List of candidate OBP genes in *L. glycinivorella*. Table S5: List of candidate OR genes in *L. glycinivorella*. Table S6: List of candidate GR genes in *L. glycinivorella*. Table S7: List of candidate IR genes in *L. glycinivorella*. Table S8: List of candidate SNMP genes in *L. glycinivorella*. Figure S1. BI phylogenetic tree of ORs from *L. glycinivorella* and representative lepidopteran species based on CDS sequences. Figure S2. BI phylogenetic tree of GRs from *L. glycinivorella* and representative lepidopteran species based on CDS sequences. Figure S3. BI phylogenetic tree of IRs from *L. glycinivorella* and representative lepidopteran species based on CDS sequences. Figure S4. BI phylogenetic tree of OBPs from *L. glycinivorella* and representative lepidopteran species based on CDS sequences. Figure S5. BI phylogenetic tree of CSPs from *L. glycinivorell* and representative lepidopteran species based on CDS sequences. Figure S6. BI phylogenetic tree of SNMPs from *L. glycinivorell* and representative lepidopteran species based on CDS sequences.

Author Contributions: Conceptualization, Y.Z. and J.S.; Investigation (sample collection), Q.W.; Phylogenetic analysis, Z.D. and R.L.; Bioinformatics and statistical analyses, J.S.; Writing—original draft, J.S.; Writing—review and editing, Y.Z., J.S. and Y.G.; Supervision, Y.Z. and S.S. Project administration, S.S. and Y.Z.; Funding acquisition, S.S. and Y.Z. All authors have read and agreed to the published version of the manuscript.

 https://doi.org/10.3390/biology15060505

Funding: This research was funded by the National Key Research and Development Program of China (Grant No. 2023YFD1401000) and the Earmarked Fund for China Agriculture Research System of MOF and MARA (Grant No. CARS-04) and the project of State Key Laboratory of Green Pesticide, Guizhou Medical University (No. GPLKF202512).

Institutional Review Board Statement: Not applicable.

Informed Consent Statement: Not applicable.

Data Availability Statement: The raw transcriptome sequencing data supporting this study have been deposited in the NCBI Sequence Read Archive (SRA) under BioProject accession number PRJNA1434332. The corresponding BioSample and SRA run accession numbers are provided in Supplementary Table S1. Additional data supporting the findings of this study are available within the article and its Supplementary Materials.

Conflicts of Interest: The authors declare no conflicts of interest.

References

1. Yang, M.; Wang, Y.; Dai, P.; Feng, D.; Hughes, A.C.; Li, H.; Zhang, A. Sympatric diversity pattern driven by the secondary contact of two deeply divergent lineages of the soybean pod borer *Leguminivora glycinivorella*. *Integr. Zool.* **2025**, *20*, 1149–1168. [CrossRef]
2. Cui, J.; Qiao, F.; Qin, B.; Xu, J.; Zhao, L.; Shi, S. Oviposition selectivity and larval fitness of soybean pod borer, *Leguminivora glycinivorella* (Lepidoptera: Olethreutidae), to different soybean varieties. *Chin. J. Oil Crop Sci.* **2023**, *45*, 592–599. [CrossRef]
3. Hu, D.; Yang, X.; Feng, J.; Zhang, X. Advances in the research and application of sex pheromone of soybean pod borer, *Leguminivora glycinivorella* (Matsumura). *Chin. J. Pestic. Sci.* **2014**, *16*, 235–244. [CrossRef]
4. Yang, M.; Wang, Z.; Wang, R.; Zhang, X.; Li, M.; Xin, J.; Qin, Y.; Zhang, C.; Meng, F. Transcriptomic and proteomic analyses of the mechanisms of overwintering diapause in soybean pod borer (*Leguminivora glycinivorella*). *Pest Manag. Sci.* **2020**, *76*, 4248–4257. [CrossRef]
5. Fang, Q.; Cao, Y.; Oo, T.H.; Zhang, C.; Yang, M.; Tang, Y.; Wang, M.; Zhang, W.; Zhang, L.; Zheng, Y.; et al. Overexpression of cry1c enhances resistance against soybean pod borer (*Leguminivora glycinivorella*) in soybean. *Plants* **2024**, *13*, 630. [CrossRef]
6. Fei, H.; Cui, J.; Zhu, S.; Xia, Y.; Xing, Y.; Gao, Y.; Shi, S. Integrative analyses of transcriptomics and metabolomics in immune response of *Leguminivora glycinivorella* Mats to *Beauveria bassiana* infection. *Insects* **2024**, *15*, 126. [CrossRef] [PubMed]
7. Xue, J.-Z.; Tariq, T.; Shen, Z.; Zhang, Y.-H.; Tang, L.-D.; Luo, R.-B.; Sun, Y.; Hu, C.-C.; Zang, L.-S. Eri silkworm eggs as a superior factitious host for mass rearing *Trichogramma leucaniae*, the key natural enemy of soybean pod borer. *Biol. Control* **2025**, *209*, 105860. [CrossRef]
8. Yoshimura, H.; Tabuchi, K.; Konishi, K. Ecological aspect of the larval parasitoid *Diadegma hiraii* (Hymenoptera: Ichneumonidae) as a potential biological control agent of soybean pod borer *Leguminivora glycinivorella* (Lepidoptera: Tortricidae). *Environ. Entomol.* **2023**, *52*, 309–316. [CrossRef]
9. Pei, T.; Cui, X.; Shi, S.; Gao, Y. An introductory review on the common brown leafhopper (*Orosius orientalis*): A new soybean pest. *Oil Crop Sci.* **2024**, *9*, 198–203. [CrossRef]
10. Leal, W.S. Odorant reception in insects: Functional and evolutionary perspectives. *Annu. Rev. Entomol.* **2026**, *71*, 275–297. [CrossRef] [PubMed]
11. Wang, B.; Jacquin-Joly, E.; Wang, G. The role of (E)-β-farnesene in tritrophic interactions: Biosynthesis, chemoreception, and evolution. *Annu. Rev. Entomol.* **2025**, *70*, 243–262. [CrossRef]
12. Agnihotri, A.R.; Roy, A.A.; Joshi, R.S. Gustatory receptors in Lepidoptera: Chemosensation and beyond. *Insect Mol. Biol.* **2016**, *25*, 519–529. [CrossRef]
13. Wicher, D.; Miazzi, F. Functional properties of insect olfactory receptors: Ionotropic receptors and odorant receptors. *Cell Tissue Res.* **2021**, *383*, 7–19. [CrossRef] [PubMed]
14. Wang, Y.; Qu, X.; Tian, Z.; Zhou, H.; Yu, Z.; Zhou, Y.; Ren, B. Molecular basis of camphor repellency in *Hyphantria cunea*. *Pestic. Biochem. Physiol.* **2024**, *204*, 106069. [CrossRef]
15. Wang, Y.; Qiu, L.; Wang, B.; Guan, Z.; Tan, Y.; Zhao, Q.; Xu, P.; Guo, H.; Jin, Y.; Zhang, J.; et al. Structural basis for odorant recognition of the insect odorant receptor OR-Orco heterocomplex. *Science* **2024**, *384*, 1453–1460. [CrossRef]
16. Zhao, J.; Chen, A.Q.; Ryu, J.; Del Mármol, J. Structural basis of odor sensing by insect heteromeric odorant receptors. *Science* **2024**, *384*, 1460–1467. [CrossRef]
17. Ai, D.; Dong, C.; Yang, B.; Yu, C.; Wang, G. A fructose receptor gene influences development and feed intake in *Helicoverpa armigera*. *Insect Sci.* **2022**, *29*, 993–1005. [CrossRef]

 https://doi.org/10.3390/biology15060505

18. Yin, N.-N.; Nuo, S.-M.; Xiao, H.-Y.; Zhao, Y.-J.; Zhu, J.-Y.; Liu, N.-Y. The ionotropic receptor gene family in Lepidoptera and Trichoptera: Annotation, evolutionary and functional perspectives. *Genomics* **2021**, *113*, 601–612. [CrossRef]
19. Ai, M.; Min, S.; Grosjean, Y.; Leblanc, C.; Bell, R.; Benton, R.; Suh, G.S.B. Acid sensing by the Drosophila olfactory system. *Nature* **2010**, *468*, 691–695. [CrossRef]
20. Silbering, A.F.; Rytz, R.; Grosjean, Y.; Abuin, L.; Ramdya, P.; Jefferis, G.S.; Benton, R. Complementary function and integrated wiring of the evolutionarily distinct *Drosophila* olfactory subsystems. *J. Neurosci.* **2011**, *31*, 13357–13375. [CrossRef] [PubMed]
21. Croset, V.; Rytz, R.; Cummins, S.F.; Budd, A.; Brawand, D.; Kaessmann, H.; Gibson, T.J.; Benton, R. Ancient protostome origin of chemosensory ionotropic glutamate receptors and the evolution of insect taste and olfaction. *PLoS Genet.* **2010**, *6*, e1001064. [CrossRef]
22. Min, S.; Ai, M.; Shin, S.A.; Suh, G.S.B. Dedicated olfactory neurons mediating attraction behavior to ammonia and amines in *Drosophila. Proc. Natl. Acad. Sci. USA* **2013**, *110*, E1321–E1329. [CrossRef]
23. Abuin, L.; Bargeton, B.; Ulbrich, M.H.; Isacoff, E.Y.; Kellenberger, S.; Benton, R. Functional architecture of olfactory ionotropic glutamate receptors. *Neuron* **2011**, *69*, 44–60. [CrossRef]
24. Budelli, G.; Ni, L.; Berciu, C.; van Giesen, L.; Knecht, Z.A.; Chang, E.C.; Kaminski, B.; Silbering, A.F.; Samuel, A.; Klein, M.; et al. Ionotropic receptors specify the morphogenesis of phasic sensors controlling rapid thermal preference in *Drosophila. Neuron* **2019**, *101*, 738–747.e3. [CrossRef]
25. Knecht, Z.A.; Silbering, A.F.; Ni, L.; Klein, M.; Budelli, G.; Bell, R.; Abuin, L.; Ferrer, A.J.; Samuel, A.; Benton, R.; et al. Distinct combinations of variant ionotropic glutamate receptors mediate thermosensation and hygrosensation in *Drosophila. eLife* **2016**, *5*, e17879. [CrossRef]
26. Enjin, A.; Zaharieva, E.E.; Frank, D.D.; Mansourian, S.; Suh, G.S.; Gallio, M.; Stensmyr, M.C. Humidity sensing in *Drosophila. Curr. Biol.* **2016**, *26*, 1352–1358. [CrossRef] [PubMed]
27. Ni, L.; Klein, M.; Svec, K.V.; Budelli, G.; Chang, E.C.; Ferrer, A.J.; Benton, R.; Samuel, A.D.T.; Garrity, P.A. The ionotropic receptors IR21a and IR25a mediate cool sensing in *Drosophila. eLife* **2016**, *5*, e13254. [CrossRef] [PubMed]
28. Koh, T.W.; He, Z.; Gorur-Shandilya, S.; Menuz, K.; Larter, N.K.; Stewart, S.; Carlson, J.R. The *Drosophila* IR20a clade of ionotropic receptors are candidate taste and pheromone receptors. *Neuron* **2014**, *83*, 850–865. [CrossRef]
29. Stewart, S.; Koh, T.W.; Ghosh, A.C.; Carlson, J.R. Candidate ionotropic taste receptors in the *Drosophila larva. Proc. Natl. Acad. Sci. USA* **2015**, *112*, 4195–4201. [CrossRef] [PubMed]
30. Tauber, J.M.; Brown, E.B.; Li, Y.; Yurgel, M.E.; Masek, P.; Keene, A.C. A subset of sweet-sensing neurons identified by IR56d are necessary and sufficient for fatty acid taste. *PLoS Genet.* **2017**, *13*, e1007059. [CrossRef]
31. Olivier, V.; Monsempes, C.; François, M.-C.; Poivet, E.; Jacquin-Joly, E. Candidate chemosensory ionotropic receptors in a Lepidoptera. *Insect Mol. Biol.* **2011**, *20*, 189–199. [CrossRef]
32. Peng, Y.; Wu, S.; Hu, S.; Wang, P.; Liu, T.; Fan, Y.; Wang, J.; Jiang, H. Ionotropic receptor 8a (Ir8a) plays an important role in acetic acid perception in the oriental fruit fly, *Bactrocera dorsalis. J. Agric. Food Chem.* **2024**, *72*, 24207–24218. [CrossRef]
33. Li, Z.; Zhang, Y.; An, X.; Wang, Q.; Khashaveh, A.; Gu, S.; Liu, S.; Zhang, Y. Identification of leg chemosensory genes and sensilla in *Apolygus lucorum. Front. Physiol.* **2020**, *11*, 276. [CrossRef]
34. Li, F.; Venthur, H.; Lin, K.; Zhang, C.; Chen, Z.; Zhou, J.-J. Insect chemosensory proteins as targets in insecticide resistance and development. *New Plant Prot.* **2025**, *2*, e70008. [CrossRef]
35. Cassau, S.; Krieger, J. The role of SNMPs in insect olfaction. *Cell Tissue Res.* **2021**, *383*, 21–33. [CrossRef] [PubMed]
36. Liu, Y.-L.; Guo, H.; Huang, L.-Q.; Pelosi, P.; Wang, C.-Z. Unique function of a chemosensory protein in the proboscis of two Helicoverpa species. *J. Exp. Biol.* **2014**, *217*, 1821–1826. [CrossRef] [PubMed]
37. Wang, H.; Zhao, R.; Gao, J.; Xiao, X.; Yin, X.; Hu, S.; Zhang, Y.; Liang, P.; Gu, S. Two cuticle-enriched chemosensory proteins confer multi-insecticide resistance in *Spodoptera frugiperda. Int. J. Biol. Macromol.* **2024**, *266*, 130941. [CrossRef] [PubMed]
38. Wang, Y.; Dong, H.; Qu, Y.; Zhou, Y.; Qin, J.; Li, K.; Luo, C.; Ren, B.; Cao, Y.; Zhang, S.; et al. Circabidian rhythm of sex pheromone reception in a scarab beetle. *Curr. Biol.* **2024**, *34*, 568–578.e5. [CrossRef]
39. Zhou, Y.; Zhou, L.; Li, Q.; Zhu, X.; Yu, Z.; Ke, H.; Chen, Q.; Ren, B. Transcriptome analysis and identification of genes related to environmental adaptation of *Grylloprimevala jilina* Zhou & Ren 2023. *Ecol. Evol.* **2023**, *13*, e10717. [CrossRef]
40. Kim, D.; Langmead, B.; Salzberg, S.L. HISAT: A fast spliced aligner with low memory requirements. *Nat. Methods* **2015**, *12*, 357–360. [CrossRef]
41. Pertea, M.; Pertea, G.M.; Antonescu, C.M.; Chang, T.-C.; Mendell, J.T.; Salzberg, S.L. StringTie enables improved reconstruction of a transcriptome from RNA-seq reads. *Nat. Biotechnol.* **2015**, *33*, 290–295. [CrossRef] [PubMed]
42. Buchfink, B.; Xie, C.; Huson, D.H. Fast and sensitive protein alignment using DIAMOND. *Nat. Methods* **2015**, *12*, 59–60. [CrossRef] [PubMed]
43. Jones, P.; Binns, D.; Chang, H.-Y.; Fraser, M.; Li, W.; McAnulla, C.; McWilliam, H.; Maslen, J.; Mitchell, A.; Nuka, G.; et al. InterProScan 5: Genome-scale protein function classification. *Bioinformatics* **2014**, *30*, 1236–1240. [CrossRef]

 https://doi.org/10.3390/biology15060505

44. Finn, R.D.; Bateman, A.; Clements, J.; Coggill, P.; Eberhardt, R.Y.; Eddy, S.R.; Heger, A.; Hetherington, K.; Holm, L.; Mistry, J.; et al. Pfam: The protein families database. *Nucleic Acids Res.* **2014**, *42*, D222–D230. [CrossRef]
45. Katoh, K.; Standley, D.M. MAFFT multiple sequence alignment software version 7: Improvements in performance and usability. *Mol. Biol. Evol.* **2013**, *30*, 772–780. [CrossRef]
46. Nguyen, L.-T.; Schmidt, H.A.; von Haeseler, A.; Minh, B.Q. IQ-TREE: A fast and effective stochastic algorithm for estimating maximum-likelihood phylogenies. *Mol. Biol. Evol.* **2015**, *32*, 268–274. [CrossRef]
47. Kalyaanamoorthy, S.; Minh, B.Q.; Wong, T.K.F.; von Haeseler, A.; Jermiin, L.S. ModelFinder: Fast model selection for accurate phylogenetic estimates. *Nat. Methods* **2017**, *14*, 587–589. [CrossRef] [PubMed]
48. Hoang, D.T.; Chernomor, O.; von Haeseler, A.; Minh, B.Q.; Vinh, L.S. UFBoot2: Improving the ultrafast bootstrap approximation. *Mol. Biol. Evol.* **2018**, *35*, 518–522. [CrossRef]
49. Ranwez, V.; Harispe, S.; Delsuc, F.; Douzery, E.J.P. MACSE: Multiple alignment of coding sequences accounting for frameshifts and stop codons. *PLoS ONE* **2011**, *6*, e22594. [CrossRef]
50. Tamura, K.; Stecher, G.; Kumar, S. MEGA11: Molecular Evolutionary Genetics Analysis version 11. *Mol. Biol. Evol.* **2021**, *38*, 3022–3027. [CrossRef]
51. Ronquist, F.; Teslenko, M.; van der Mark, P.; Ayres, D.L.; Darling, A.; Höhna, S.; Larget, B.; Liu, L.; Suchard, M.A.; Huelsenbeck, J.P. MrBayes 3.2: Efficient Bayesian phylogenetic inference and model choice across a large model space. *Syst. Biol.* **2012**, *61*, 539–542. [CrossRef] [PubMed]
52. Zhang, D.; Gao, F.; Jakovlić, I.; Zou, H.; Zhang, J.; Li, W.X.; Wang, G.T. PhyloSuite: An integrated and scalable desktop platform for streamlined molecular sequence data management and evolutionary phylogenetics studies. *Mol. Ecol. Resour.* **2020**, *20*, 348–355. [CrossRef]
53. Letunic, I.; Bork, P. Interactive Tree Of Life (iTOL) v5: An online tool for phylogenetic tree display and annotation. *Nucleic Acids Res.* **2021**, *49*, W293–W296. [CrossRef]
54. Li, G.; Du, J.; Li, Y.; Wu, J. Identification of putative olfactory genes from the oriental fruit moth *Grapholita molesta* via an antennal transcriptome analysis. *PLoS ONE* **2015**, *10*, e0142193. [CrossRef]
55. Guo, M.; Du, L.; Chen, Q.; Feng, Y.; Zhang, J.; Zhang, X.; Tian, K.; Cao, S.; Huang, T.; Jacquin-Joly, E.; et al. Odorant receptors for detecting flowering plant cues are functionally conserved across moths and butterflies. *Mol. Biol. Evol.* **2021**, *38*, 1413–1427. [CrossRef]
56. Cao, D.; Liu, Y.; Wei, J.; Liao, X.; Walker, W.B., III; Li, J.; Wang, G. Identification of candidate olfactory genes in *Chilo suppressalis* by antennal transcriptome analysis. *Int. J. Biol. Sci.* **2014**, *10*, 846–860. [CrossRef] [PubMed]
57. Jacquin-Joly, E.; Legeai, F.; Montagné, N.; Monsempes, C.; François, M.-C.; Poulain, J.; Gavory, F.; Walker, W.B., III; Hansson, B.S.; Larsson, M.C. Candidate chemosensory genes in female antennae of the noctuid moth *Spodoptera littoralis*. *Int. J. Biol. Sci.* **2012**, *8*, 1036–1050. [CrossRef] [PubMed]
58. Walker, W.B., III; Gonzalez, F.; Garczynski, S.F.; Witzgall, P. The chemosensory receptors of codling moth *Cydia pomonella*—Expression in larvae and adults. *Sci. Rep.* **2016**, *6*, 23518. [CrossRef] [PubMed]
59. Yang, J.; Guo, H.; Jiang, N.-L.; Tang, R.; Li, G.-C.; Huang, L.-Q.; van Loon, J.J.A.; Wang, C.-Z. Identification of a gustatory receptor tuned to sinigrin in the cabbage butterfly *Pieris rapae*. *PLoS Genet.* **2021**, *17*, e1009527. [CrossRef]
60. Koenig, C.; Hirsh, A.; Bucks, S.; Klinner, C.; Vogel, H.; Shukla, A.; Mansfield, J.H.; Morton, B.; Hansson, B.S.; Grosse-Wilde, E. A reference gene set for chemosensory receptor genes of *Manduca sexta*. *Insect Biochem. Mol. Biol.* **2015**, *66*, 51–63. [CrossRef]
61. Guo, H.; Cheng, T.; Chen, Z.; Jiang, L.; Guo, Y.; Liu, J.; Li, S.; Taniai, K.; Asaoka, K.; Kadono-Okuda, K.; et al. Expression map of a complete set of gustatory receptor genes in chemosensory organs of *Bombyx mori*. *Insect Biochem. Mol. Biol.* **2017**, *82*, 74–82. [CrossRef]
62. Briscoe, A.D.; Macias-Muñoz, A.; Kozak, K.M.; Walters, J.R.; Yuan, F.; Jamie, G.A.; Martin, S.H.; Dasmahapatra, K.K.; Ferguson, L.C.; Mallet, J.; et al. Female behaviour drives expression and evolution of gustatory receptors in butterflies. *PLoS Genet.* **2013**, *9*, e1003620. [CrossRef]
63. Xu, W.; Papanicolaou, A.; Zhang, H.-J.; Anderson, A. Expansion of a bitter taste receptor family in the polyphagous insect *Helicoverpa armigera*. *Sci. Rep.* **2016**, *6*, 23666. [CrossRef]
64. Dong, J.-F.; Yang, H.-B.; Li, D.-X.; Yu, H.-Q.; Tian, C.-H. Identification and expression analysis of chemosensory receptors in the tarsi of fall armyworm, *Spodoptera frugiperda* (J. E. Smith). *Front. Physiol.* **2023**, *14*, 1177297. [CrossRef]
65. Zhan, S.; Merlin, C.; Boore, J.L.; Reppert, S.M. The monarch butterfly genome yields insights into long-distance migration. *Cell* **2011**, *147*, 1171–1185. [CrossRef]
66. Corcoran, J.A.; Jordan, M.D.; Thrimawithana, A.H.; Crowhurst, R.N.; Newcomb, R.D. The peripheral olfactory repertoire of the lightbrown apple moth, *Epiphyas postvittana*. *PLoS ONE* **2015**, *10*, e0128596. [CrossRef]
67. Li, H.; Hong, X.; Zeng, F.; Bai, C. Identification and expression profiles of olfactory-related genes based on transcriptome analysis in *Plodia interpunctella* (Lepidoptera: Pyralidae). *Arch. Insect Biochem. Physiol.* **2023**, *114*, e22061. [CrossRef] [PubMed]

68. Huang, C.; Zhang, X.; He, D.; Wu, Q.; Tang, R.; Xing, L.; Liu, W.; Wang, W.; Liu, B.; Xi, Y.; et al. Comparative Genomics Provide Insights into Function and Evolution of Odorant Binding Proteins in *Cydia pomonella*. *Front. Physiol.* **2021**, *12*, 690185. [CrossRef] [PubMed]
69. Chang, X.-Q.; Nie, X.-P.; Zhang, Z.; Zeng, F.-F.; Lv, L.; Zhang, S.; Wang, M.-Q. De novo analysis of the oriental armyworm *Mythimna separata* antennal transcriptome and expression patterns of odorant-binding proteins. *Comp. Biochem. Physiol. Part D Genom. Proteom.* **2017**, *22*, 120–130. [CrossRef] [PubMed]
70. Huang, X.; Yang, J.; Zou, J.; Wen, X.; Wu, T.; Tian, X.; Luo, J.; Niu, Y.; Huang, X. Identification of odorant binding protein and chemosensory protein genes in *Protegira songi* (Lepidoptera: Noctuidae) via transcriptome analysis. *J. Asia-Pac. Entomol.* **2024**, *27*, 102255. [CrossRef]
71. Agnihotri, A.; Liu, N.; Xu, W. Chemosensory Proteins (CSPs) in the Cotton Bollworm *Helicoverpa armigera*. *Insects* **2022**, *13*, 29. [CrossRef]
72. Liu, S.; Chang, H.; Liu, W.; Cui, W.; Liu, Y.; Wang, Y.; Ren, B.; Wang, G. Essential role for SNMP1 in detection of sex pheromones in *Helicoverpa armigera*. *Insect Biochem. Mol. Biol.* **2020**, *127*, 103485. [CrossRef]
73. Venthur, H.; Zhou, J.-J. Odorant receptors and odorant-binding proteins as insect pest control targets: A comparative analysis. *Front. Physiol.* **2018**, *9*, 1163. [CrossRef]
74. Li, R.; Jiang, G.-F.; Shu, X.-H.; Wang, Y.-Q.; Li, M.-J. Identification and Expression Profile Analysis of Chemosensory Genes From the Antennal Transcriptome of Bamboo Locust (*Ceracris kiangsu*). *Front. Physiol.* **2020**, *11*, 889. [CrossRef] [PubMed]
75. Liu, Q.; Liu, W.; Zeng, B.-S.; Wang, G.; Hao, D.; Huang, Y. Deletion of the *Bombyx mori* odorant receptor co-receptor (BmOrco) impairs olfactory sensitivity in silkworms. *Insect Biochem. Mol. Biol.* **2017**, *86*, 58–67. [CrossRef] [PubMed]
76. Sakurai, T.; Nakagawa, T.; Mitsuno, H.; Mori, H.; Endo, Y.; Tanoue, S.; Yasukochi, Y.; Touhara, K.; Nishioka, T. Identification and functional characterization of a sex pheromone receptor in the silkmoth *Bombyx mori*. *Proc. Natl. Acad. Sci. USA* **2004**, *101*, 16653–16658. [CrossRef]
77. Zhang, D.-D.; Löfstedt, C. Moth pheromone receptors: Gene sequences, function, and evolution. *Front. Ecol. Evol.* **2015**, *3*, 105. [CrossRef]
78. Bengtsson, J.M.; Gonzalez, F.; Cattaneo, A.M.; Montagné, N.; Walker, W.B.; Bengtsson, M.; Anfora, G.; Ignell, R.; Jacquin-Joly, E.; Witzgall, P. A predicted sex pheromone receptor of codling moth *Cydia pomonella* detects the plant volatile pear ester. *Front. Ecol. Evol.* **2014**, *2*, 33. [CrossRef]
79. Li, R.-T.; Huang, L.-Q.; Dong, J.-F.; Wang, C.-Z. A moth odorant receptor highly expressed in the ovipositor is involved in detecting host-plant volatiles. *eLife* **2020**, *9*, e53706. [CrossRef]
80. Miyamoto, T.; Slone, J.; Song, X.; Amrein, H. A fructose receptor functions as a nutrient sensor in the *Drosophila* brain. *Cell* **2012**, *151*, 1113–1125. [CrossRef]
81. Jiang, X.-J.; Ning, C.; Guo, H.; Jia, Y.-Y.; Huang, L.-Q.; Qu, M.-J.; Wang, C.-Z. A gustatory receptor tuned to D-fructose in antennal sensilla chaetica of *Helicoverpa armigera*. *Insect Biochem. Mol. Biol.* **2015**, *60*, 39–46. [CrossRef] [PubMed]
82. Xu, W.; Zhang, H.-J.; Anderson, A. A sugar gustatory receptor identified from the foregut of cotton bollworm *Helicoverpa armigera*. *J. Chem. Ecol.* **2012**, *38*, 1513–1520. [CrossRef] [PubMed]
83. Sato, R. Molecular Functions and Physiological Roles of Gustatory Receptors of the Silkworm *Bombyx mori*. *Int. J. Mol. Sci.* **2024**, *25*, 10157. [CrossRef]
84. Ning, C.; Yang, K.; Xu, M.; Huang, L.-Q.; Wang, C.-Z. Functional validation of the carbon dioxide receptor in labial palps of *Helicoverpa armigera* moths. *Insect Biochem. Mol. Biol.* **2016**, *73*, 12–19. [CrossRef]
85. Zhang, J.; Duan, S.; Wang, W.; Liu, D.; Wang, Y. Molecular Basis of CO_2 Sensing in *Hyphantria cunea*. *Int. J. Mol. Sci.* **2024**, *25*, 5987. [CrossRef] [PubMed]
86. Ai, M.; Blais, S.; Park, J.-Y.; Min, S.; Neubert, T.A.; Suh, G.S.B. Ionotropic glutamate receptors IR64a and IR8a form a functional odorant receptor complex in vivo in Drosophila. *J. Neurosci.* **2013**, *33*, 10741–10749. [CrossRef]
87. Raji, J.I.; Melo, N.; Castillo, J.S.; Gonzalez, S.; Saldana, V.; Stensmyr, M.C.; DeGennaro, M. Aedes aegypti mosquitoes detect acidic volatiles found in human odor using the IR8a pathway. *Curr. Biol.* **2019**, *29*, 1253–1262.e7. [CrossRef]
88. Zhang, X.-X.; Yang, B.; Sun, D.-D.; Guo, M.-B.; Zhang, J.; Wang, G.-R. Ionotropic receptor 8a is involved in the attraction of *Helicoverpa armigera* to acetic acid. *Insect Sci.* **2022**, *29*, 657–668. [CrossRef]
89. Pelosi, P.; Iovinella, I.; Felicioli, A.; Dani, F.R. Soluble proteins of chemical communication: An overview across arthropods. *Front. Physiol.* **2014**, *5*, 320. [CrossRef]
90. Zheng, Z.-C.; Li, D.-Z.; Zhou, A.; Yi, S.-C.; Liu, H.; Wang, M.-Q. Predicted structure of a Minus-C OBP from *Batocera horsfieldi* (Hope) suggests an intermediate structure in evolution of OBPs. *Sci. Rep.* **2016**, *6*, 33981. [CrossRef]
91. Xiao, S.; Sun, J.S.; Carlson, J.R. Robust olfactory responses in the absence of odorant binding proteins. *eLife* **2019**, *8*, e51040. [CrossRef]
92. Tu, J.; Wang, Z.; Yang, F.; Liu, H.; Qiao, G.; Zhang, A.; Wang, S. The female-biased general odorant binding protein 2 of *Semiothisa cinerearia* displays binding affinity for biologically active host plant volatiles. *Biology* **2024**, *13*, 274. [CrossRef]

 https://doi.org/10.3390/biology15060505

93. Pelosi, P.; Iovinella, I.; Zhu, J.; Wang, G.; Dani, F.R. Beyond chemoreception: Diverse tasks of soluble olfactory proteins in insects. *Biol. Rev.* **2018**, *93*, 184–200. [CrossRef] [PubMed]
94. Yi, X.; Qi, J.; Zhou, X.; Hu, M.-Y.; Zhong, G.-H. Differential expression of chemosensory-protein genes in midguts in response to diet of *Spodoptera litura*. *Sci. Rep.* **2017**, *7*, 296. [CrossRef] [PubMed]
95. Khuhro, S.A.; Yan, Q.; Liao, H.; Zhu, G.-H.; Sun, J.-B.; Dong, S.-L. Expression profile and functional characterization suggesting the involvement of three chemosensory proteins in perception of host plant volatiles in *Chilo suppressalis*. *J. Insect Sci.* **2018**, *18*, 6. [CrossRef]
96. Tsouri, A.; Douris, V. The role of chemosensory proteins in insecticide resistance: A review. *Insects* **2025**, *16*, 496. [CrossRef] [PubMed]
97. Cassau, S.; Krieger, J. Evidence for a role of SNMP2 and antennal support cells in sensillum lymph clearance processes of moth pheromone-responsive sensilla. *Insect Biochem. Mol. Biol.* **2024**, *164*, 104046. [CrossRef]
98. Forstner, M.; Gohl, T.; Gondesen, I.; Raming, K.; Breer, H.; Krieger, J. Differential expression of SNMP-1 and SNMP-2 proteins in pheromone-sensitive hairs of moths. *Chem. Senses* **2008**, *33*, 291–299. [CrossRef]

https://doi.org/10.3390/biology15060505

biology

MDPI

Article

A Species-Specific *COI* PCR Approach for Discriminating Co-Occurring Thrips Species Using Crude DNA Extracts

Qingxuan Qiao [1,2,†], Yaqiong Chen [1,2,†], Jing Chen [2], Ting Chen [1], Huiting Feng [2], Yussuf Mohamed Salum [2], Han Wang [2], Lu Tang [2], Hongrui Zhang [3], Zheng Chen [1], Tao Lin [1], Hui Wei [1,*] and Weiyi He [2,*]

1 Fujian Engineering Research Center for Green Pest Management, Fujian Key Laboratory for Monitoring and Integrated Management of Crop Pests, Institute of Plant Protection, Fujian Academy of Agricultural Sciences, Fuzhou 350013, China; qqx17862665489@163.com (Q.Q.); chen_yaqiong@126.com (Y.C.); ct318815@163.com (T.C.); chenzheng@faas.cn (Z.C.); maludongzuo@163.com (T.L.)
2 State Key Laboratory of Agricultural and Forestry Biosecurity, Institute of Applied Ecology, Fujian Agriculture and Forestry University, Fuzhou 350002, China; cj13276011790@163.com (J.C.); feng06092022@126.com (H.F.); 13919892472@163.com (H.W.); tlttll2003@163.com (L.T.)
3 Plant Protection College, Yunnan Agricultural University, Kunming 650201, China; hongruizh@126.com
* Correspondence: weihui@faas.cn (H.W.); wy.he@fafu.edu.cn (W.H.)
† These authors contributed equally to this work.

Simple Summary

Thrips are tiny insects that feed on many crops and can spread plant viruses. Different thrips species often co-occur on the same host plants but are morphologically similar, making species-level molecular identification challenging under field-oriented conditions. This is important because species differ in how much damage they cause; how they respond to insecticides; and whether they spread viruses. In this study, we developed a simple detection system that can quickly tell apart four important thrips species that attack crops in southern China; including one globally invasive species; the western flower thrips. The test uses species-diagnostic regions of the mitochondrial *COI* gene and a short-turnaround DNA extraction workflow for processing individual insects. Our results show that species-specific and reproducible amplification can be achieved under mixed-species backgrounds, and that PCR inhibition can be effectively alleviated by mild dilution of crude insect lysates. This framework supports laboratory- and field-oriented molecular identification workflows for greenhouse and open-field thrips surveillance and provides a methodological basis for potential integration with simplified visual nucleic acid readout platforms in future applications.

Abstract

Thrips are cosmopolitan agricultural pests and important vectors of plant viruses, and the increasing coexistence of multiple morphologically similar species has intensified the demand for species-specific molecular identification. However, traditional morphological identification and PCR assays using universal primers are often inadequate for mixed-species samples and field-adaptable application. In this study, we developed a species-specific molecular identification framework targeting a polymorphism-rich region of the mitochondrial cytochrome c oxidase subunit I (*COI*) gene, which is more time-efficient than sequencing-based *COI* DNA barcoding, for four economically important thrips species in southern China, including the globally invasive *Frankliniella occidentalis*. By aligning *COI* sequences, polymorphism-rich regions were identified and used to design four species-specific primer pairs, each containing a diagnostic 3′-terminal nucleotide. These primers were combined with a PBS-based DNA extraction workflow optimized for single-insect samples that minimizes dependence on column-based purification. The assay achieved a practical detection limit of 1 ng per reaction, demonstrated species-specific amplification, and maintained reproducible amplification at DNA inputs of $\geq$1 ng per reaction. Notably, PCR inhibition caused by crude extracts was effectively alleviated by fivefold dilution. Although the chemical identities of the inhibitors remain unknown, interspecific variation in inhibition strength was observed, with *T. hawaiiensis* exhibiting the strongest suppression, possibly due to differences in lysate composition. This integrated framework balances target specificity, operational simplicity, and dilution-mitigated inhibition, providing a field-adaptable tool for thrips species identification and invasive species monitoring. Moreover, it provides a species-specific molecular foundation for downstream integration with visual nucleic acid detection platforms, such as the CRISPR/Cas12a system, thereby facilitating the future development of portable molecular identification workflows for small agricultural pests.

Keywords: *COI* gene; species-specific primers; PBS-based DNA extraction

1. Introduction

Polyphagous thrips are globally important agricultural pests widely distributed across tropical and subtropical regions. Their minute body size and high reproductive capacity enable infestation across diverse crops [1,2]. Thrips possess piercing-sucking mouthparts and initiate feeding by rasping plant surfaces before inserting their stylets to extract cellular contents. This feeding behavior induces tissue necrosis, leaf deformation, and fruit developmental abnormalities, ultimately compromising both crop yield and quality [3,4]. Furthermore, several species function as competent vectors of plant viruses such as tomato spotted wilt virus (TSWV), impatiens necrotic spot virus (INSV), and groundnut ringspot virus (GRSV), thereby posing a dual threat through both direct feeding damage and virus transmission [5,6]. Together, these traits render thrips a persistent threat to the biosecurity and sustainability of modern agricultural production.

In protected cropping systems, thrips exhibit continuous year-round occurrence and rapid generational turnover [7,8]. Their capacity for parthenogenetic reproduction, rapid evolution of insecticide resistance, and pronounced cryptic behavior further contributes to their persistence in high-value horticultural crops [8,9]. In southern coastal regions of China, the expansion of protected cultivation and intensified cross-regional trade have markedly reshaped the regional distribution patterns and dominant species composition of thrips populations [10,11]. Currently, the flower thrips *Frankliniella intonsa* (Trybom, 1895) (Thysanoptera: Thripidae), the bean thrips *Megalurothrips usitatus* (Bagnall, 1913), and the Hawaiian flower thrips *Thrips hawaiiensis* (Morgan, 1913) dominate local agricultural systems in Fujian Province [12,13], whereas the invasive western flower thrips *Frankliniella occidentalis* (Pergande, 1895), originally absent from most provinces in China, has in recent years rapidly expanded its distribution range, progressively establishing stable populations and exhibiting strong competitive displacement of native species, thereby emerging as a dominant pest in greenhouse-based protected cropping systems [14].

As a result, agricultural ecosystems increasingly exhibit the coexistence of multiple thrips species, with interspecific resource partitioning and competitive interactions that may influence population dynamics [15,16]. Accurate identification and real-time monitoring of dominant thrips populations are therefore fundamental for elucidating population succession and coexistence mechanisms and serve as essential prerequisites for precision pest management [17,18]. However, the high morphological similarity among species, fragility of specimens, and frequent cooccurrence substantially limit the accuracy and

 https://doi.org/10.3390/biology15020171

efficiency of traditional morphological approaches, particularly under field monitoring conditions [19,20]. Although mitochondrial cytochrome c oxidase subunit I (*COI*) markers have been widely applied in molecular identification, barcoding approaches rely on sequencing platforms and thus fall short of supporting rapid, field-adaptable diagnostics [21]. Despite the broad adoption for amplifying conserved mitochondrial barcoding loci, universal *COI* primers are not intended for direct species discrimination at the PCR level. When applied to samples containing closely related or co-occurring thrips species, universal primers may yield mixed amplification products, which require subsequent sequencing for accurate species assignment and thus limit their suitability for rapid, field-adaptable identification without sequencing [22]. The limitations are further amplified by the practical constraints of current molecular workflows, which usually depend on specialized laboratory equipment and controlled laboratory conditions. These include thermocyclers, centrifuges, and gel-based detection systems, as well as multi-step DNA purification procedure, that involve multiple handling and transfer steps, even in commonly used column-based extraction workflows, thereby limiting the applicability to low-input analyses from individual insects under field conditions [23]. Compounding this challenge, crude lysates from individual insects may contain co-extracted substances that interfere with downstream PCR amplification, thereby reducing amplification efficiency and reliability rather than the efficiency of DNA extraction itself [24,25].

In summary, the frequent coexistence of multiple thrips species and dynamic shifts in species dominance underscores the pressing need for a molecular species identification framework that integrates species-level resolution, operational simplicity, and field-adaptable applicability. To address this challenge, the present study focused on thrips species most frequently encountered in local agricultural systems, together with a globally invasive species *F. occidentalis*, and performed a systematic assessment of mitochondrial *COI* polymorphisms to pinpoint hypervariable regions conducive to species-specific primer development. Concurrently, six DNA extraction protocols were evaluated and optimized to establish a short-turnaround, species-specific molecular identification that reconciles species-level resolution with field-compatible applicability and is more time-efficient than sequencing-based barcoding. Departing from conventional *COI* barcoding approaches restricted to laboratory-based sequencing, this study combines polymorphism-guided primer design with a PBS-based DNA extraction strategy for single-insect samples that requires no column-based purification steps, establishing a modular identification framework oriented toward laboratory and field applications. Such integration effectively bridges the gap between laboratory molecular assays and field-adaptable surveillance and lays a molecular foundation for seamless incorporation of isothermal amplification and visual nucleic acid detection platforms, including CRISPR/Cas12a-based assays, to advance intelligent pest monitoring and to accelerate the transition toward portable, data-driven molecular species identification frameworks in smart agriculture.

2. Materials and Methods

2.1. Insect Materials and Rearing Maintenance

Four thrips species used in this study were *F. intonsa*, *F. occidentalis*, *M. usitatus*, and *T. hawaiiensis*. Among them, *F. intonsa*, *M. usitatus*, and *T. hawaiiensis* are dominant native species commonly found in agricultural systems throughout Fujian Province, while *F. occidentalis* has rapidly spread in China. The *F. occidentalis* laboratory strain was established in 2018 from a starter population provided by the Institute of Biotechnology and Germplasm Resources, Yunnan Academy of Agricultural Sciences. Field-derived strains of *F. intonsa* (2021, collected on cucumber, *Cucumis sativus* L., 1753), *M. usitatus* (2021, collected on cowpea, *Vigna unguiculata* (L.) Walp., 1843), and *T. hawaiiensis* (2017, collected on sweet

 https://doi.org/10.3390/biology15020171

osmanthus, *Osmanthus fragrans* (Thunb.) Lour., 1790) were established independently from populations sampled in Minhou, Fuzhou, Fujian, China. All four species have been maintained under laboratory conditions for multiple years. Each species has been reared independently under strict biosecurity protocols to prevent cross-contamination and to ensure stable laboratory populations. Species identity was regularly verified using morphological characteristics and mitochondrial *COI* barcoding. All species were maintained in a controlled-environment chamber (CRX-350B, Fujian Zhongyouling Technology Co., Ltd., Fuzhou, China) set at 22 ± 1 °C, 75 ± 5% RH, and a 16:8 h light:dark (L:D) photoperiod. Strain health and developmental synchrony were routinely monitored throughout rearing.

Rearing followed a previously described system [26], which was further optimized in this study to enhance egg collection efficiency and minimize contamination risk. For *F. intonsa*, *F. occidentalis*, and *T. hawaiiensis*, adults were maintained and allowed to oviposit in custom frustum-shaped polypropylene cages (Cage A; top diameter 8 cm, bottom diameter 11 cm, height 8 cm). The cage top was covered with breathable mesh, and the bottom was sealed with Parafilm® M (Bemis Company, Inc., Sheboygan Falls, WI, USA) and positioned above a water-filled tray to maintain membrane moisture while preventing direct water contact. Approximately 300 healthy adults (mixed sexes) were introduced into each cage (≥3 cages per species per batch), and germinated broad beans (*Vicia faba*) were supplied as food and replaced every 48 h to prevent microbial contamination. Females oviposited through the Parafilm into the membrane–water interface. Eggs were gently collected by rinsing the membrane three times with 10 mL of distilled water per rinse and subsequently vacuum-filtered using a SHZ-D(III) water-circulating vacuum pump (Zhengzhou Greatwall, Zhengzhou, China) onto black Whatman No. 1 filter paper (Cytiva, Marlborough, MA, USA) or black cotton cloth to enhance visibility. By contrast, *M. usitatus* showed poor adaptation to the Parafilm-based oviposition setup; therefore, germinated broad beans were placed directly inside the cage as the oviposition substrate. After 48 h, beans containing *M. usitatus* eggs and eggs of the other species collected from the Parafilm membranes were each transferred to their respective rectangular rearing cages (Cage B; 12 × 8 × 6 cm; fine-mesh sides) for hatching and nymphal development. A moistened filter paper sheet (Whatman No. 1, Cytiva, Marlborough, MA, USA) was placed on the cage bottom to maintain humidity and provide a pupation substrate, and both food and filter paper were replaced every 2 days. All thrips strains were maintained under the same controlled rearing conditions described above. Empty rearing cages and associated tools were sprayed with 70% ethanol and disinfected with UV light for 30 min before introducing the next generation. Developmental progress was monitored daily, and temperature and humidity were recorded using the built-in sensors of the environmental chamber to ensure stable rearing conditions. Individuals were collected at the designated developmental stages for subsequent analyses.

2.2. Morphometric Analysis and COI Reference Sequence Generation in Four Thrips Species

To ensure consistency across replicates and to eliminate variability caused by sexual dimorphism, ten adult females of each species were collected from laboratory-reared populations at 1–3 days post-eclosion. All individuals were fully sclerotized, morphologically intact, and free from visible deformities. Because pronounced sexual dimorphism in body size and external morphology exists between males and females, only females were examined for morphometric analysis to eliminate sex-related variability and ensure consistent and comparable measurements. This restriction was applied exclusively to the morphological measurements. Individual specimens were positioned at the center of a microscope slide for stereomicroscopic imaging. The wings were symmetrically aligned along the body axis in the same focal plane to minimize distortion caused by improper posture.

https://doi.org/10.3390/biology15020171

Stereomicroscope imaging was performed using ZSA402 stereomicroscope (Chongqing Optec Instrument Co., Ltd., Chongqing, China). All images were calibrated using embedded 200 μm scale bars, and morphometric measurements were performed in millimeters (mm) using the S-EYE 2.0 image analysis system integrated with the stereomicroscope (Chongqing Optec Instrument Co., Ltd., Chongqing, China). All measurements were recorded to two decimal places.

All individuals used in this study were directly sampled from four well-established laboratory strains whose identity had been routinely verified through morphological examination and mitochondrial *COI* barcoding. At the start of the experimental workflow, species identity was further confirmed based on full-length mitochondrial *COI* sequence information prior to subsequent method development. For molecular characterization and generation of high-quality *COI* reference sequences, genomic DNA was extracted from 50 adult individuals randomly selected from laboratory-maintained strains, including both females and males, using a conventional buffer-based DNA extraction method (CB-Std), which consists of Tris–EDTA–NaCl buffer preparation, SDS-mediated lysis, and alcohol precipitation. Full-length mitochondrial *COI* fragments (approximately 1500 bp) from these individuals were then amplified and sequenced to generate high-quality templates for primer design. The extraction buffer consisted of 1.461 g NaCl (Solarbio, Beijing, China), 5 mL of 1 M Tris-HCl (Solarbio, Beijing, China), and 5 mL of 0.5 M EDTA (Solarbio, Beijing, China), which was adjusted to pH 8.0 and brought to a final volume of 50 mL with sterile ultrapure water (Milli-Q system; MilliporeSigma, Burlington, MA, USA). The buffer was stored at 4 °C. Each specimen was homogenized in a 1.5 mL microcentrifuge tube using a sterile micro-pestle. For lysis, 500 μL of extraction buffer and 100 μL of 10% SDS (Solarbio, Beijing, China) were added to each tube, followed by incubation at 56 °C for 1 h. Subsequently, 200 μL of 2 M potassium acetate (Solarbio, Beijing, China) was added, and the mixture was incubated at 4 °C for 1 h. After centrifugation at 14,000 rpm for 10 min, the supernatant was collected and mixed with an equal volume of pre-chilled isopropanol (Macklin Biochemical Co., Ltd., Shanghai, China; AR grade) to precipitate DNA. The resulting DNA pellet was air-dried, resuspended in 20 μL of sterile ultrapure water (Milli-Q system; MilliporeSigma, Burlington, MA, USA), and stored at −20 °C until use.

PCR amplification was carried out in a 25 μL reaction volume containing 12.5 μL of 2× Taq PCR Master Mix (Vazyme, Nanjing, China), 1 μL each of forward and reverse primers (10 μM), 1 μL of genomic DNA template, and nuclease-free water to a final volume of 25 μL. Thermal cycling conditions were as follows: initial denaturation at 95 °C for 3 min, followed by 35 cycles of 95 °C for 15 s, 58 °C for 15 s, and 72 °C for 30 s, with a final extension at 72 °C for 10 min. PCR products were verified by electrophoresis on a 2% agarose gel, and target fragments were excised and purified using a Gel Extraction Kit (Omega Bio-Tek, Norcross, GA, USA). Purified PCR fragments were ligated into the pJET1.2 vector using the CloneJET PCR Cloning Kit (Thermo Fisher Scientific, Waltham, MA, USA) and transformed into *Escherichia coli* DH5α competent cells (Tiangen Biotech Co., Ltd., Beijing, China). Transformed cells were plated on LB agar supplemented with ampicillin (100 μg/mL) and incubated at 37 °C for 12–16 h. Positive colonies were inoculated into LB broth containing ampicillin and incubated at 37 °C with shaking at 200 rpm for 6 h prior to plasmid extraction. Inserts were confirmed by PCR using pJET1.2 universal primers, and the products were validated by 1% agarose gel electrophoresis. Sequencing was carried out in both directions by Sangon Biotech (Shanghai, China). Raw sequence reads were assembled and manually curated using SnapGene (version 6.2.1; GSL Biotech LLC, San Diego, CA, USA; https://www.snapgene.com/), generating high-quality *COI* reference sequences for the four target thrips species. These verified sequences were subsequently used as templates for species-specific primer design and downstream analyses.

 https://doi.org/10.3390/biology15020171

2.3. Sequence Alignment and Species-Specific Primer Design

Based on the obtained *COI* sequences, multiple sequence alignment and characterization were performed using MEGA software (version 12.0; Arizona State University, Tempe, AZ, USA). Given that *COI* is a protein-coding gene with a conserved reading frame, sequence alignment was primarily inspected and refined manually. Nucleotide composition (A%, T%, G%, and C%) and GC content were calculated to assess base composition bias and sequence complexity across the four thrips species. Pairwise genetic distances were estimated under the *p*-distance model, with missing data treated using pairwise deletion, resulting in a genetic distance matrix among species. To identify species-specific regions, the aligned *COI* sequences were further visualized and manually examined using SnapGene (version 6.2.1; GSL Biotech LLC, San Diego, CA, USA; https://www.snapgene.com/) to locate variable nucleotide sites. Regions exhibiting high polymorphism were annotated as potential targets for primer design.

During primer design, highly conserved regions were excluded to avoid reduced discriminatory power. Candidate primers were selected from the most variable *COI* segments, with priority given to those containing at least one species-specific nucleotide at the 3′ terminus. Design principles were as follows: (1) targeting regions with the highest nucleotide variability; (2) maintaining amplicon lengths between 600 and 800 bp; (3) avoiding regions with high homology to non-target species; and (4) minimizing the formation of hairpins, dimers, or self-complementarity. All primers were designed using Primer Premier (version 6.0; PREMIER Biosoft, Palo Alto, CA, USA; https://www.premierbiosoft.com/) and further checked in SnapGene (version 6.2.1; GSL Biotech LLC, San Diego, CA, USA; https://www.snapgene.com/) to confirm complete matching to the intended target region and to exclude predicted secondary structures in the primers. Final primer sequences are provided in Table S1.

2.4. Validation of Primer Specificity and Performance

Amplification efficiency and species specificity were evaluated by performing PCR assays using genomic DNA from the four thrips species under the same thermal cycling program as *COI* amplification. Amplicons were resolved on 1% agarose gels to verify the presence of the expected species-specific band. Each primer pair was tested against its cognate target and all non-target species. Primer pairs yielding a single, distinct band in the target species and no amplification in non-targets were considered specific and selected for further validation. Primer performance under mixed-template backgrounds was assessed using two mock DNA communities to simulate interspecific mixtures. The first mixture consisted of genomic DNA from *F. intonsa*, *F. occidentalis*, *M. usitatus*, and *T. hawaiiensis*, with each species contributing an equal amount to a total of 50 ng DNA per reaction. The second mixture contained only the three non-target species, with the total DNA amount likewise adjusted to 50 ng per reaction, excluding the target species for a given primer pair. Reactions were run under the same conditions as single species assays. Successful species discrimination was defined as amplification only in mixtures containing the target species, with no bands in target-absent mixtures.

2.5. Short-Turnaround DNA Extraction and Single Thrips Adult PCR Detection

To evaluate short-turnaround DNA extraction strategies suitable for single-insect diagnostics under resource-limited conditions, six extraction methods were systematically compared using individual adult thrips. These included a modified conventional protocol (CB-Std) and five short-turnaround alternatives designed to reduce procedural complexity while maintaining compatibility with downstream PCR.

https://doi.org/10.3390/biology15020171

The CB-Std method (Section 2.2) served as the reference protocol and involved SDS-based chemical lysis, potassium acetate precipitation, and DNA resuspension. For single-adult processing, the total reaction volume was reduced from 500 μL to 20 μL with proportional adjustment of reagents. The five short-turnaround extraction methods were classified according to buffer composition and lysis strategy. Three detergent-free methods employed sterile distilled water (SDW; Milli-Q system; MilliporeSigma, Burlington, MA, USA), phosphate-buffered saline (PBS; ABKBio, Xiamen Aibikang Biotechnology Co., Ltd., Xiamen, China), or 20 mM EDTA (Solarbio, Beijing, China) as lysis media. For these methods, individual thrips were homogenized in 20 μL of the respective buffer and subjected to brief heat lysis at 98 °C for 2 min, followed by cooling to room temperature. Importantly, the PBS-based protocol did not include SDS or any other detergents, and the resulting lysates were used directly as PCR templates without further purification. In contrast, a simplified detergent-assisted protocol (CB-SDS) utilized a reduced-concentration lysis buffer containing 0.2% SDS to evaluate the effect of mild detergent inclusion on DNA release and amplification efficiency. A nitrocellulose membrane contact transfer (NCM) method was also evaluated. In this approach, individual thrips were directly pressed onto nitrocellulose membranes (Solarbio, Beijing, China) and air-dried at room temperature. Prior to PCR, membrane fragments (approximately 0.5 × 0.5 cm) bearing the thrips contact trace were excised and transferred into 20 μL of sterile water (Milli-Q system; MilliporeSigma, Burlington, MA, USA). The thrips contact trace was defined as residual biological material transferred from the insect body surface during physical contact, including cuticular fragments, cellular debris, and trace amounts of internal tissue or hemolymph sufficient to serve as a DNA template. Samples were heated at 98 °C for 2 min to release DNA and then cooled to room temperature before PCR.

DNA concentration and purity were measured using a NanoDrop microvolume spectrophotometer (Thermo Fisher Scientific, Waltham, MA, USA). For each extraction method, ten biological replicates were tested per thrips species, resulting in a total of 240 PCR amplification reactions. Amplification performance was assessed by 1% agarose gel electrophoresis to compare band clarity, specificity, and PCR compatibility across extraction methods.

2.6. Determination of PCR Detection Sensitivity

Genomic DNA extracted from single adults of the four species was serially diluted to five concentration levels, corresponding to DNA inputs of 50 ng, 10 ng, 1 ng, 100 pg, and 10 pg per 25 μL PCR reaction. PCR amplification was conducted using the species-specific primer pairs designed for each thrips species, following the same reaction composition and thermal cycling parameters described in Section 2.2. The amplified products were resolved on 1% agarose gels, and band visibility and clarity were evaluated to assess amplification performance across DNA input levels. The lowest DNA amount consistently producing a visually distinct and reproducible band was defined as the limit of detection (LOD) for each assay. Each DNA input level was conducted in three biological replicates to confirm reproducibility and reliability.

2.7. Evaluation of Species-Specific Sensitivity Across Mixed Population Ratios

To assess the detection sensitivity of target thrips across varying abundance ratios, a gradient mixture assay was conducted. Gradient mixtures were prepared by combining one adult of the target species with one to four individuals of each of the other three species, yielding target-to-background ratios of 1:1:1:1, 1:2:2:2, 1:3:3:3, and 1:4:4:4. Corresponding negative controls were prepared using the same ratios but excluding the target species (non-target thrips only). Each mixture was subsequently homogenized in 50 μL of 1× PBS using

 https://doi.org/10.3390/biology15020171

the PBS-based short-turnaround extraction protocol, ensuring consistency among treatments. Each treatment was performed with three biological replicates, and 1 μL of extracted DNA was used per PCR reaction to maintain consistent template input across assays.

2.8. Statistical Analysis

Morphometric parameters (body length and width) and DNA extraction metrics (yield, purity, and storage stability) were expressed as mean ± standard deviation (SD) calculated from ten biological replicates ($n = 10$) per species. Statistical analyses were performed using IBM SPSS Statistics (version 26.0; IBM Corp., Armonk, NY, USA; https://www.ibm.com/products/spss-statistics (accessed on 14 January 2026)). One-way analysis of variance (ANOVA) was applied to compare means among treatments, followed by Tukey's multiple comparison test, with significance set at $p < 0.05$. GraphPad Prism (version 9.0.0; GraphPad Software, San Diego, CA, USA; https://www.graphpad.com/) was used for data visualization and figure preparation.

3. Results

3.1. External Morphology and Interspecific COI Divergence Among Four Thrips Species

Adult females of the four thrips species examined exhibit highly similar overall external morphology at the structural level under routine observation (Figure 1A–D), with no readily diagnostic structural characters distinguishable among species.

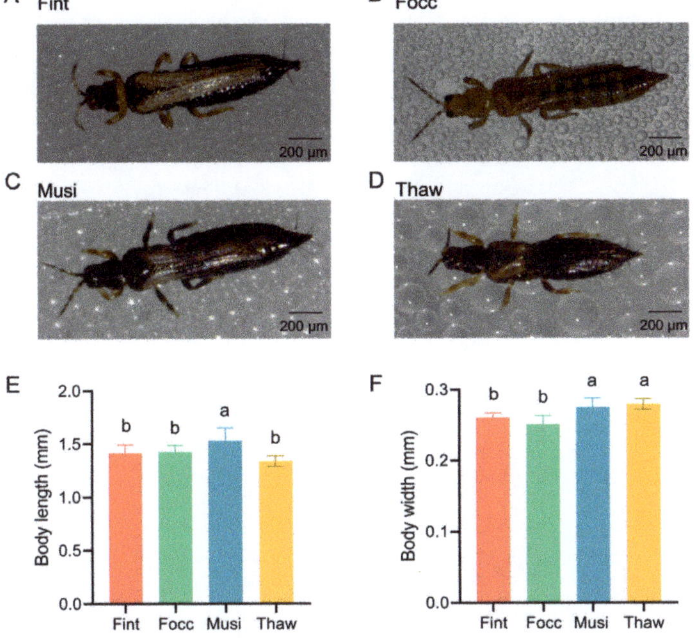

Figure 1. External morphology and body size comparison among four thrips species. (**A–D**) Representative adult females of Fint (*F. intonsa*); Focc (*F. occidentalis*); Musi (*M. usitatus*); Thaw (*T. hawaiiensis*) under stereomicroscopy. Scale bars: 200 μm. (**E**) Comparison of body length among the four species ($n = 10$ per species). (**F**) Comparison of body width among the four species ($n = 10$ per species). Bars sharing the same lowercase letter do not differ significantly, whereas bars labeled with different letters indicate statistically significant differences among species (one-way ANOVA followed by Tukey's post hoc test, $p < 0.05$). Data are presented as mean ± standard deviation (SD).

https://doi.org/10.3390/biology15020171

Simple body size measurements revealed limited but statistically significant differences in body length and width among species (Figure 1E,F). Specifically, *M. usitatus* exhibited a greater mean body length, whereas *M. usitatus* and *T. hawaiiensis* showed slightly increased body width compared with the other species (one-way ANOVA followed by Tukey's post hoc test, $p < 0.05$). However, substantial overlap was observed across species for both parameters, indicating that these measurements alone provide insufficient resolution for reliable species-level identification. Although minor differences in pigmentation intensity were visually apparent among species, such traits are highly variable and strongly influenced by age, physiological condition, and environmental factors, and therefore cannot be considered reliable diagnostic characters under field conditions.

To enable species-specific molecular identification, the mitochondrial *COI* gene was selected as the diagnostic target, and its interspecific variation among the four species was analyzed. PCR amplification of the *COI* gene produced bands migrating between the 1000 and 2000 bp markers and visually closer to the 2000 bp marker in all four thrips species (Figure 2A). Subsequent sequence analysis revealed that the amplicons were 1554 bp in *F. intonsa* and *F. occidentalis* and 1542 bp in *M. usitatus* and *T. hawaiiensis*, indicating minor interspecific length variation within the *COI* coding region that is not visually resolvable on agarose gels. Alignments to the corresponding NCBI references (*F. intonsa* NC_081965.1, *F. occidentalis* NC_018370.1, *M. usitatus* NC_070094.1, and *T. hawaiiensis* MW582621.1) showed high identity, with only a few nucleotide mismatches. Based on this verification, the experimentally derived *COI* sequences were used as templates for species-specific primer design and downstream analyses. Given the strong concordance with public entries, no additional GenBank submissions were made, and all downstream analyses were based on the experimentally derived *COI* sequences.

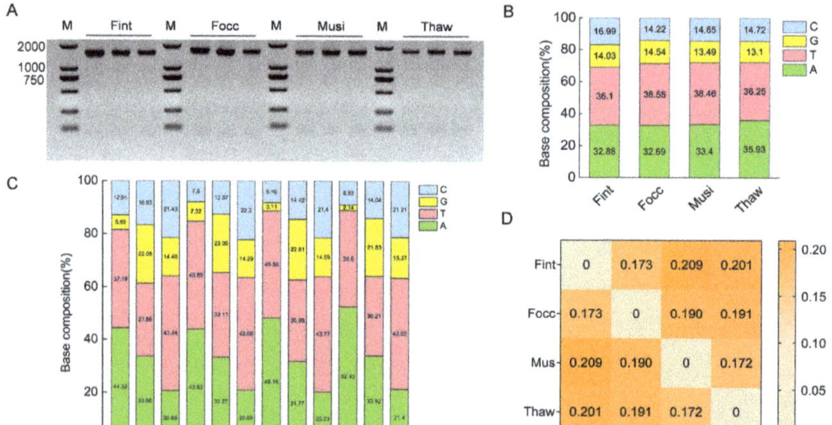

Figure 2. Sequence characterization of *COI* genes sequences in four thrips species. (**A**) Agarose gel electrophoresis showing successful amplification of full-length *COI* sequences in Fint (*F. intonsa*), Focc (*F. occidentalis*), Musi (*M. usitatus*), and Thaw (*T. hawaiiensis*). Representative amplicons of the expected size are shown for each species, confirming consistent *COI* amplification for subsequent sequencing analyses. M: DNA marker (DL2000). (**B**) Overall base composition of *COI* sequences, showing a strong AT bias in all four species. (**C**) Base composition by codon position, with pronounced AT enrichment at the first and third positions. (**D**) Pairwise genetic distances (uncorrected *p*-distance) among species, calculated based on full-length *COI* sequences.

https://doi.org/10.3390/biology15020171

Base-composition analysis revealed a pronounced AT bias across all four species, with overall A + T content exceeding 68% (Figure 2B). Among the four species, *T. hawaiiensis* exhibited the highest adenine (A) content (35.93%), whereas *F. occidentalis* and *M. usitatus* showed the highest thymine (T) levels (38.55% and 38.46%, respectively). Guanine (G) was the least abundant nucleotide, accounting for only 13.10–14.54% of total bases. Interspecific differences were most evident for adenine (A) and cytosine (C), indicating species-specific nucleotide usage preferences.

Positional analysis further demonstrated that the first and third codon positions exhibited the strongest AT enrichment, with A + T content exceeding 80% (Figure 2C). At the first codon position, *T. hawaiiensis* showed A content as high as 52.43%, whereas G and C were markedly underrepresented. G was particularly scarce at this position, with minimum values of 5.59% in *F. intonsa* and 2.14% in *T. hawaiiensis*. The second codon position displayed a more balanced nucleotide distribution, consistent with stronger functional constraints on protein-coding regions. By contrast, the third codon position was strongly enriched in thymine, reaching 43.44% in *F. intonsa*, with the other species showing similar levels (approximately 42%). Collectively, these codon-specific patterns highlight diagnostic polymorphisms within the *COI* coding region that can serve as reproducible molecular markers for species discrimination.

Pairwise uncorrected *p*-distance analysis of the *COI* coding region revealed interspecific values ranging from 0.172 to 0.209 among the four thrips species (Figure 2D). The smallest *p*-distance (0.173) was observed between *F. intonsa* and *F. occidentalis*, whereas the largest value (0.209) occurred between *M. usitatus* and *T. hawaiiensis*, representing only a modest increase relative to the minimum. Overall, all interspecific comparisons showed *p*-distance values exceeding 0.17, indicating clear sequence divergence among species in the *COI* coding region. Because *p*-distance is an uncorrected metric, these values summarize observed sequence differences and may underestimate underlying evolutionary change if multiple substitutions have occurred. Taken together, these results demonstrate consistent interspecific *COI* sequence differences among all species pairs examined, providing a suitable basis for subsequent species identification and species-specific primer design in this study.

3.2. Species-Specific COI Primer Screening and Validation

To identify regions suitable for species-specific primer design, *COI* sequences from the four thrips species were aligned and analyzed for interspecific divergence patterns as described above (Figure S1). A polymorphism-rich region spanning positions 600–1300 bp, marked by dense nucleotide substitutions and pronounced interspecific variation, was identified and selected as the target region for primer development. Within this region, three candidate primer pairs were designed for each of the four thrips species in accordance with standard primer design principles, resulting in twelve primer sets in total (Table S1).

Pooled genomic DNA extracted from 50 adult individuals per species was used as the template for primer screening to ensure experimental consistency and reproducibility. All candidate primers were first tested under identical PCR conditions using single species DNA templates to assess amplification performance and specificity, and primer pairs yielding weak or non-specific bands were excluded from further analyses (Figure S2). Subsequently, mixed-template assays containing DNA from multiple species were conducted to further evaluate potential cross-reactivity and confirm primer specificity (Figure S3). Only primer pairs that produced a single distinct amplicon in the presence of their respective target DNA, without any cross-reactive bands in non-target mixtures, were retained. After two rounds of screening under both single and mixed template conditions, one optimal primer pair was identified for each species and was subsequently renamed based on its

https://doi.org/10.3390/biology15020171

species designation. The selected primer pairs annealed to *COI* coding regions exhibiting the greatest interspecific nucleotide divergence and met all predefined design criteria. Notably, within each primer pair, both the forward and reverse primers carried at least one diagnostic nucleotide at the 3′ terminus uniquely matching the target species, thereby ensuring high specificity and robust discriminatory power (Figure 3).

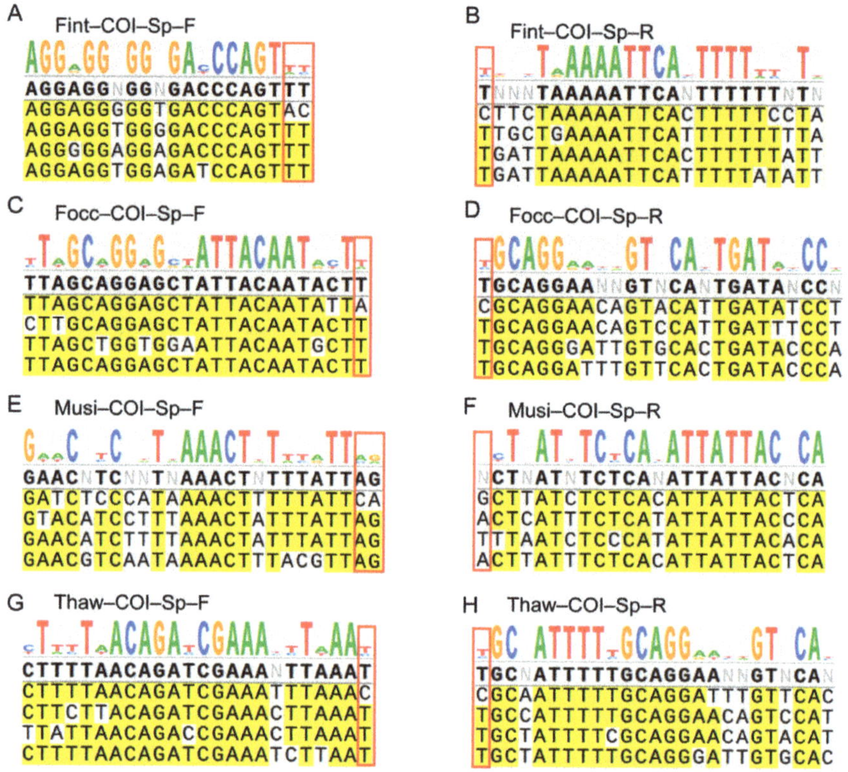

Figure 3. Primer binding regions and interspecific *COI* variation among four thrips species. (**A–H**) The upper panel shows a sequence logo summarizing nucleotide variability across aligned *COI* sequences from the four thrips species within the displayed window. The lower panel shows a multiple sequence alignment of representative *COI* sequences from the four species; the first sequence in each alignment corresponds to the species-specific primer designed for the target species indicated in the panel title, followed by the *COI* sequences of the four thrips species. Red boxes highlight the 3′-terminal region of the primer-binding site that was emphasized during primer design to ensure species specificity. The color scheme and bold formatting are automatically generated by the visualization software and do not indicate additional biological annotations.

Each optimized primer pair was further validated and produced a single, distinct amplicon of the expected size in its corresponding target species (580 bp in *F. intonsa*, 590 bp in *F. occidentalis*, 652 bp in *M. usitatus*, and 561 bp in *T. hawaiiensis*), whereas no amplification was detected in the non-target species (Figure 4).

 https://doi.org/10.3390/biology15020171

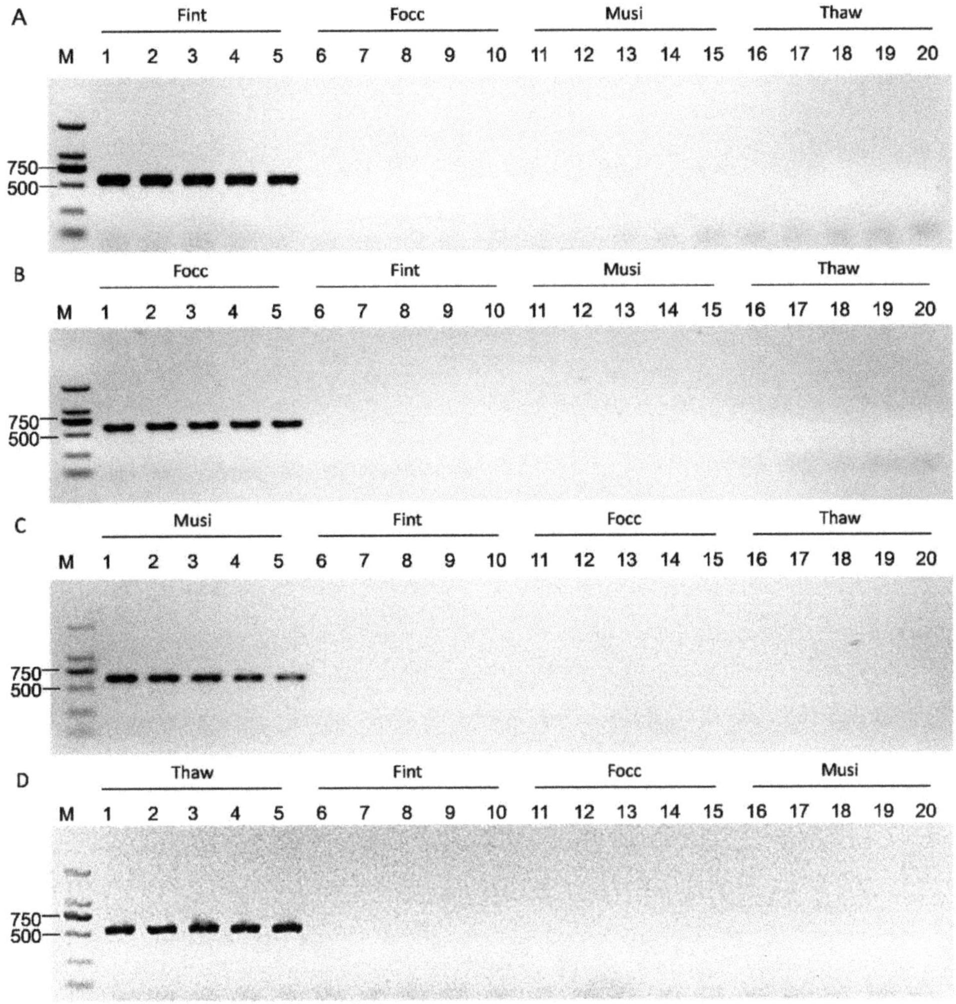

Figure 4. Species-specific PCR amplification with *COI* primers under single species template conditions. (**A**) Fint (*F. intonsa*); (**B**) Focc (*F. occidentalis*); (**C**) Musi (*M. usitatus*); (**D**) Thaw (*T. hawaiiensis*). A single, distinct amplicon was observed exclusively in its corresponding species, with sizes of 580 bp for Fint, 590 bp for Focc, 652 bp for Musi, and 561 bp for Thaw. No cross-reactivity was observed. Qualitative validation of primer specificity based on amplification presence or absence rather than visual comparison of interspecific differences. All PCR assays were independently repeated five times with consistent results. M: DNA marker (DL2000).

Subsequently, PCR assays using mixed DNA templates (a total of 50 ng DNA per reaction, with equal contributions from each species), either containing or lacking the target species, yielded clear single bands exclusively in mixtures containing the respective targets (Figure 5). These results confirm that the optimized *COI* primer system enables species-specific and reproducible discrimination of closely related thrips species under mixed-species backgrounds.

https://doi.org/10.3390/biology15020171

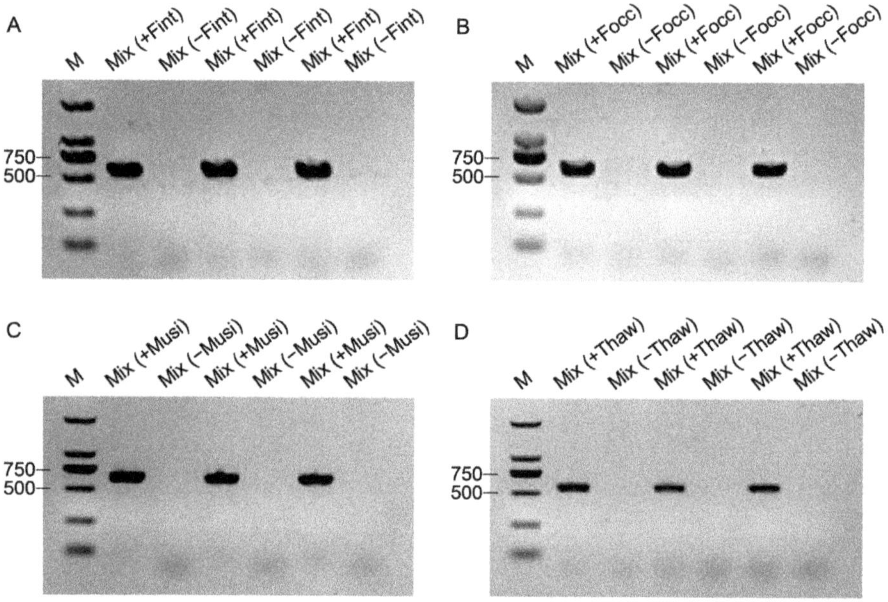

Figure 5. Species-specific PCR amplification using a mixture of single species DNA templates. (**A**) Fint (*F. intonsa*), 580 bp; (**B**) Focc (*F. occidentalis*), 590 bp; (**C**) Musi (*M. usitatus*), 652 bp; (**D**) Thaw (*T. hawaiiensis*), 561 bp. For each panel, three independent biological replicates are shown for both "+" and "−" mixtures. PCR assays were performed using species-specific primers with mixed DNA templates (total 50 ng DNA per reaction). "+": target species included; "–": excluded. Distinct amplicons were detected exclusively in "+" mixtures, with no cross-amplification observed in "–" groups. M: DNA marker (DL2000).

3.3. Optimization and Performance Assessment of DNA Extraction Methods for Single Thrips Adult Detection

Based on the establishment of species-specific *COI* primers capable of accurately distinguishing four thrips species, six DNA extraction protocols were systematically evaluated for single insect samples to enhance their field-level diagnostic utility. Accordingly, five non-instrumental lysis methods (PBS, EDTA, SDW, NCM, and CB-SDS) were evaluated, with CB-Std employed as the positive control.

Representative amplification patterns, classified as strong, moderate, and weak, are presented (Figure 6). Immediately after DNA extraction (week 0), the PBS and CB-Std groups produced distinct species-specific bands (indicated by arrowheads) across all four thrips species, yielding clear and reproducible amplicons of the expected size, with a 100% amplification success rate (40/40; Figure 6 and Figure S4). In contrast, the remaining four rapid methods exhibited markedly lower efficiency, with success rates generally below 25% and poor amplification clarity and reproducibility (Figure 6, Figure S5 and S6). Accordingly, these four methods were deemed unsuitable for downstream PCR applications, both after 1–4 weeks of refrigerated storage and for molecular detection of single thrips adult. Subsequent weekly PCR validations revealed that both PBS and CB-Std maintained clear and consistent amplification during the first two weeks. However, from week 3 onward, band intensities for *F. occidentalis* and *T. hawaiiensis* gradually weaken under both extraction methods, with one or two samples failing to produce detectable amplicons. By week 4, this trend became more pronounced, particularly in *T. hawaiiensis*, where bands became increasingly faint, although the overall amplification success rate still remained above 80% (Figures S7–S10).

https://doi.org/10.3390/biology15020171

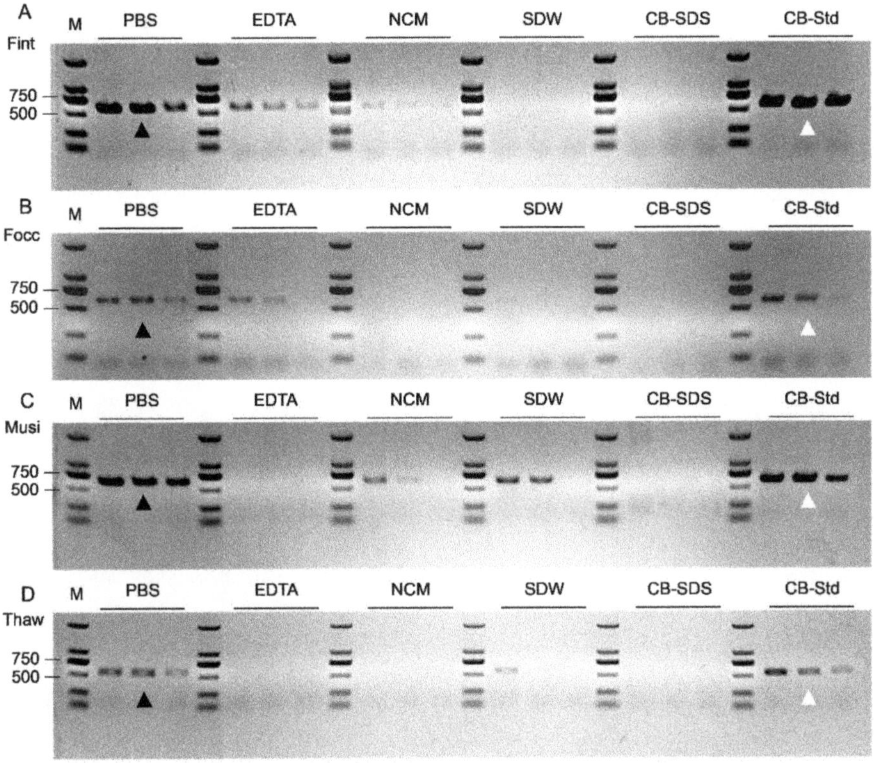

Figure 6. PCR performance of six DNA extraction methods for single thrips adult. (**A**) Fint (*F. intonsa*, 580 bp). (**B**) Focc (*F. occidentalis*, 590 bp). (**C**) Musi (*M. usitatus*, 652 bp). (**D**) Thaw (*T. hawaiiensis*, 561 bp). Week 0 PCR products of single thrips adult extracted using PBS, EDTA, NCM, SDW, CB-SDS, and CB-Std are shown. Black triangles indicate PBS as the optimal short-turnaround method, and white triangles indicate CB-Std as the standard control. M: DNA marker (DL2000).

To assess DNA preservation under refrigerated conditions, DNA concentration and purity were monitored from week 0 to week 4 (Figure 7). At week 0, CB-SDS yielded the highest DNA concentrations, ranging from 194.68 ± 38.52 ng/μL (*T. hawaiiensis*) to 290.41 ± 133.13 ng/μL (*F. occidentalis*). EDTA yielded concentrations between 111.52 ± 44.85 ng/μL (*T. hawaiiensis*) and 204.11 ± 144.56 ng/μL (*M. usitatus*), whereas SDW ranged from 87.84 ± 26.17 ng/μL (*T. hawaiiensis*) to 209.69 ± 48.64 ng/μL (*F. intonsa*), both representing intermediate levels. PBS produced moderate but highly reproducible concentrations (80.61–108.42 ng/μL, with SDs mostly below 25 ng/μL except in *T. hawaiiensis*), indicating good consistency across species. NCM yielded DNA concentrations ranging from 35.81 ± 7.37 ng/μL in *F. occidentalis* to 72.15 ± 14.05 ng/μL in *M. usitatus*, while CB-Std ranged from 34.00 ± 4.66 ng/μL in *T. hawaiiensis* to 61.50 ± 16.11 ng/μL in *F. intonsa*. During the four-week storage period, DNA concentrations obtained using PBS remained above 69.60 ng/μL. By contrast, CB-Std consistently yielded low concentrations (approximately 33.46–61.50 ng/μL) with minimal variation, indicating high reproducibility despite limited DNA recovery. CB-SDS, however, exhibited a marked decline in DNA concentration starting at week 1, accompanied by increased variability among samples. EDTA and SDW showed a more gradual downward trend over the same period, while NCM consistently produced the lowest concentrations, indicating limited potential for long-term storage.

https://doi.org/10.3390/biology15020171

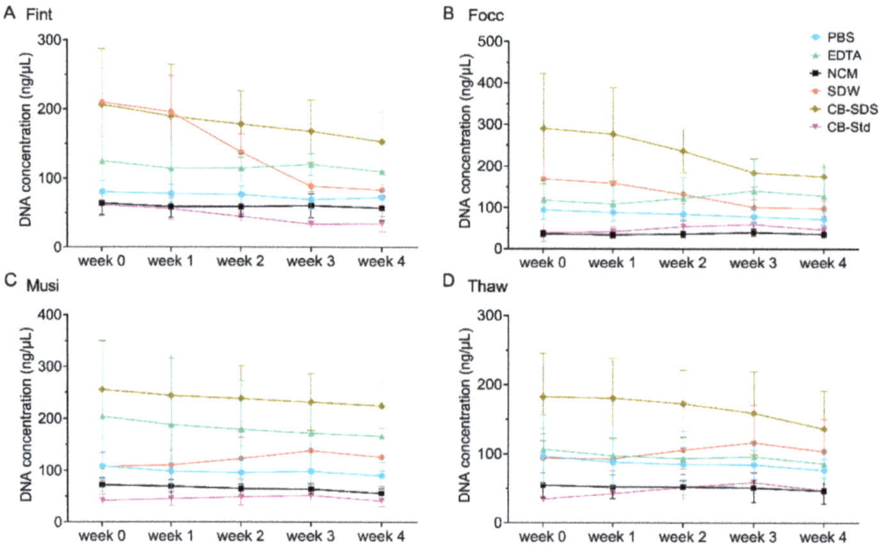

Figure 7. DNA concentration dynamics of four thrips species under refrigerated storage based on six extraction methods. (**A**) Fint (*F. intonsa*); (**B**) Focc (*F. occidentalis*); (**C**) Musi (*M. usitatus*); (**D**) Thaw (*T. hawaiiensis*). DNA concentrations (ng/µL) were measured weekly over a four-week period with six extraction buffers: PBS, EDTA, NCM, SDW, CB-SDS, and CB-Std. Data represent mean ± standard deviation (SD) from ten biological replicates.

Analysis of A260/A280 and A260/A230 ratios revealed method-specific temporal trends in DNA absorbance profiles over time (Figures S11 and S12). At week 0, CB-Std showed the highest A260/A280 ratios (ranging from 2.22 ± 0.07 in *T. hawaiiensis* to 2.49 ± 0.20 in *F. occidentalis*) but consistently low A260/A230 values (ranging 0.29 ± 0.33 in *F. intonsa* to 0.60 ± 0.30 in *T. hawaiiensis*), suggesting RNA carryover and residual organic or chaotropic contaminants. In contrast, PBS exhibited the most balanced absorbance profile, with A260/A280 ratios ranging from 1.41 ± 0.11 (*F. occidentalis*) to 1.58 ± 0.14 (*M. usitatus*) and A260/A230 ratios ranging from 0.59 ± 0.05 (*F. intonsa*) to 0.67 ± 0.10 (*F. occidentalis*), reflecting relatively stable UV absorbance characteristics of the crude extracts. Throughout the storage period, CB-Std maintained high A260/A280 ratios (approximately 2.2–2.5); however, A260/A230 ratios showed a progressive decline in most species, dropping below 0.20 by week 4 in *F. occidentalis*, *M. usitatus*, and *T. hawaiiensis*, suggesting co-extraction of residual impurities and potential DNA degradation. By contrast, PBS maintained both ratios within relatively narrow ranges, with A260/A280 values between 1.39 and 1.64 and A260/A230 values between 0.54 and 0.73, showing minimal week-to-week variation (SD < 0.15) and demonstrating superior stability of absorbance profiles over the storage period. NCM displayed a similar yet slightly lower profile, with moderate and stable values across both ratios. EDTA consistently produced moderately low A260/A280 ratios (approximately 1.18–1.62) and uniformly low A260/A230 ratios (below 0.32) across all species, consistent with interference from residual chelating agents. SDW and CB-SDS yielded the lowest and most variable values for both ratios, particularly after week 2, suggesting compromised DNA quality and possible accumulation of inhibitory substances. Taken together, PBS exhibited the highest stability and reproducibility in DNA absorbance ratios and DNA yield across extraction methods and species, highlighting its utility as a reproducible and field-compatible buffer for single insect molecular identification.

https://doi.org/10.3390/biology15020171

3.4. Validation of Detection Sensitivity for the Species-Specific PCR System

To determine the detection sensitivity of the PCR assay, PBS-extracted genomic DNA was tested at five DNA input levels (50, 10, 1, 0.1, and 0.01 ng per 25 μL reaction) across four thrips species, with three biological replicates per species. All primer pairs consistently yielded single, species-specific amplicons of the expected sizes at DNA inputs ≥ 1 ng, with no off-target amplification observed (Figure 8A–D). Collectively, these results define 1 ng per reaction as the LOD for specific and reproducible amplification. The sporadic amplification at 0.1 ng in *F. intonsa* and *M. usitatus* delineates a practical detection window of 0.1–1 ng for single thrips adult. Minor interspecific differences were observed at low template levels, reflecting slight variation in amplification efficiency. Importantly, these differences did not compromise the overall diagnostic performance of the assay.

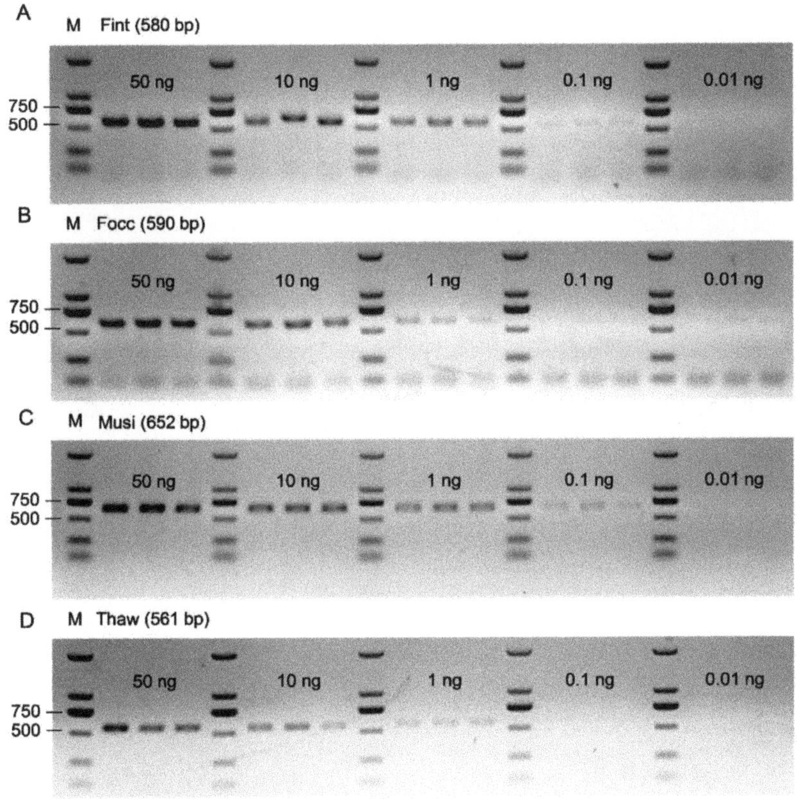

Figure 8. Detection sensitivity of the species-specific PCR assay in four thrips species. (**A–D**) Amplification profiles of Fint (*F. intonsa*, 580 bp), Focc (*F. occidentalis*, 590 bp), Musi (*M. usitatus*, 652 bp), and Thaw (*T. hawaiiensis*, 561 bp). PCR was performed using PBS-extracted genomic DNA at five DNA input levels (50, 10, 1, 0.1, and 0.01 ng per 25 μL; 1 μL input). Clear, single, species-specific bands were consistently observed at ≥1 ng, with faint or sporadic bands at 0.1 ng in Fint and Musi, and no detectable bands at 0.01 ng. M, DNA marker (DL2000).

3.5. Effect of Species Background and Sample Input on PCR Inhibition and Its Alleviation by Dilution

To evaluate species-specific amplification under mixed-species backgrounds, one adult of the target thrips was combined with one to four individuals of each of the three non-target species, forming target-to-background ratios of 1:1:1:1, 1:2:2:2, 1:3:3:3, and 1:4:4:4.

 https://doi.org/10.3390/biology15020171

Genomic DNA from each mixture was extracted with PBS. In undiluted reactions, *F. intonsa* and *M. usitatus* consistently yielded distinct, species-specific amplicons across all ratios. By contrast, *F. occidentalis* and *T. hawaiiensis* were detectable only at the ratio of 1:1:1:1, and their signals progressively attenuated and eventually disappeared as the proportion of non-target individuals increased. Following fivefold dilution, specific amplification was restored for all four species irrespective of mixing ratio (Figure 9). This recovery suggests that PCR inhibition, probably resulting from impurities or template competition in crude lysates, was effectively mitigated by dilution. Consistent results from three biological replicates indicate that the assay is reproducible and suitable for detecting targets at low abundance within complex and field-relevant backgrounds.

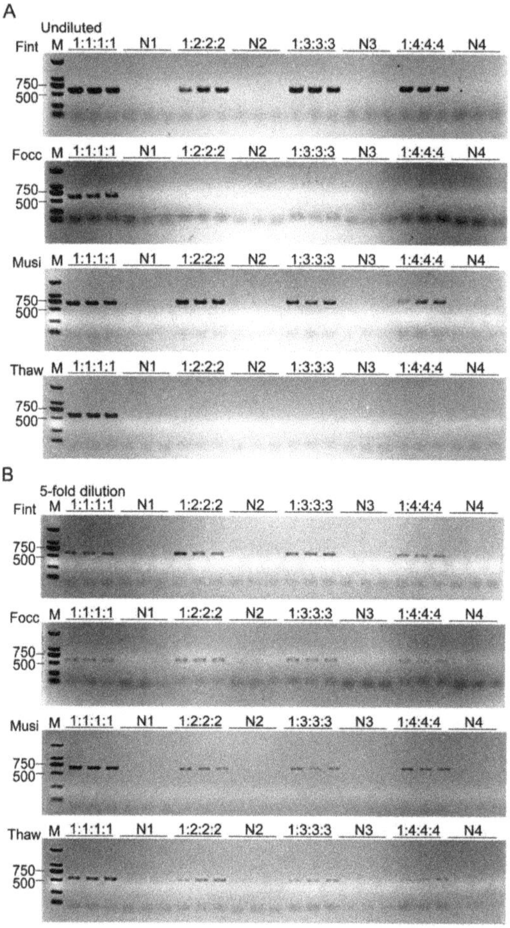

Figure 9. Species-specific amplification of the PCR assay under mixed-species background interference. (**A**) Undiluted lysates. (**B**) Fivefold-diluted lysates. Each lane contains PBS-extracted genomic DNA from a mixture of one individual of the target species (*F. intonsa*, *F. occidentalis*, *M. usitatus*, or *T. hawaiiensis*) and one to four individuals of each of the three non-target species, forming target-to-background ratios of 1:1:1:1, 1:2:2:2, 1:3:3:3, and 1:4:4:4. M: DNA marker (DL2000); N1–N4, mixtures containing only non-target species at the same ratios.

To assess interspecific differences in lysate-induced effects on amplification efficiency, single species abundance-gradient assays were conducted using PBS-prepared crude

https://doi.org/10.3390/biology15020171

lysates. Crude lysates from *F. intonsa*, *F. occidentalis*, and *M. usitatus* consistently produced clear amplicons across samples containing one to four individuals, and these results remained unchanged after fivefold dilution, indicating relatively weak inhibitory effects. In contrast, *T. hawaiiensis* exhibited a strong, non-linear inhibition pattern, with markedly reduced amplification in two-individual samples, where bands became nearly undetectable. Signals were slightly recovered at higher inputs, suggesting that inhibition intensity varied with sample composition. After uniform fivefold dilution, the overall amplification pattern remained similar, indicating that dilution had limited effects on relieving inhibition in *T. hawaiiensis* lysates (Figure 10). These findings highlight pronounced interspecific variation in lysate-induced inhibition, with *T. hawaiiensis* showing the strongest and most persistent suppression among the four species, consistent with a higher abundance of PCR inhibitors in its crude lysates.

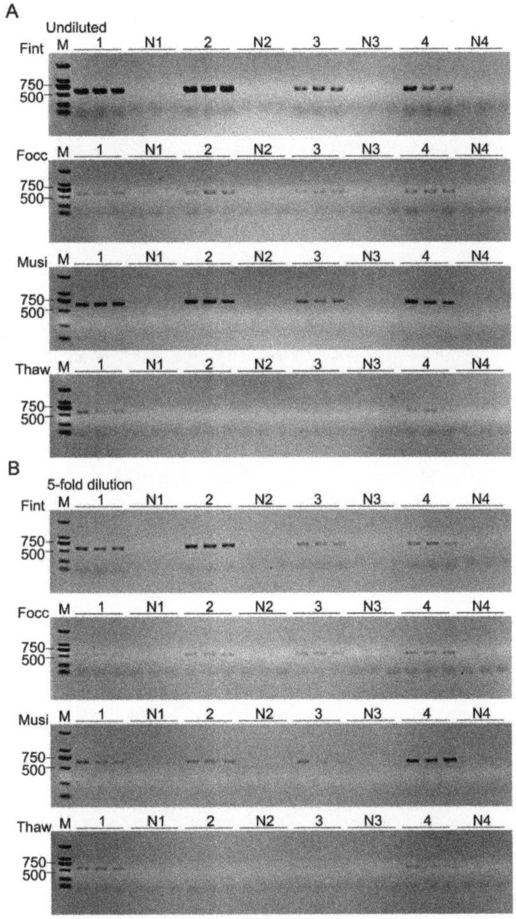

Figure 10. Interspecific variation in PCR inhibition. (**A**) Undiluted lysates. (**B**) Fivefold-diluted lysates. Samples were prepared by pooling 1–4 individuals of *Fint* (*F. intonsa*), *Focc* (*F. occidentalis*), *Musi* (*M. usitatus*), or *Thaw* (*T. hawaiiensis*). PBS-extracted crude DNA was used as the template. 1–4, number of pooled individuals; N1–N4, no-template negative controls corresponding to each gradient. M, DNA marker (DL2000).

https://doi.org/10.3390/biology15020171

4. Discussion

The high taxonomic diversity and pronounced morphological similarity of thrips communities limit the reliability of conventional morphology-based identification, particularly when specimens are fragile or collected at variable developmental stages [27–29]. To address these practical limitations, we established a species-specific molecular identification framework targeting the mitochondrial *COI* region, which enables reproducible, target-specific amplification in both individual and mixed-species samples. Unlike conventional DNA barcoding approaches that rely on universal primers targeting conserved regions, followed by post-PCR sequencing and subsequent sequence alignment [30,31], our system employs a polymorphism-informed, 3′-terminal discriminatory primer design anchored in species-specific polymorphic sites, enabling selective amplification in combination with a PBS-based short-turnaround DNA extraction protocol that minimizes procedural complexity and enhances field compatibility. The resulting framework integrates a balance among analytical specificity, operational simplicity, and dilution-mitigated inhibition, providing a species-specific and reproducible molecular identification framework that is time-efficient compared with sequencing-based barcoding for thrips species of agricultural importance.

The mitochondrial *COI* gene, characterized by a combination of conserved and hypervariable regions, remains the most widely used molecular marker for animal species identification owing to its robust interspecific resolution [32,33]. In this study, alignment of *COI* sequences revealed polymorphism-enriched regions that supported species-specific discrimination while maintaining amplification efficiency. Although conventional DNA barcoding approaches based on universal primers and sequence alignment have greatly facilitated reference library development, they frequently suffer from cross-species amplification and ambiguous results in complex, mixed-species samples [34,35]. To overcome these limitations, we employed a polymorphism-guided, primer-level engineering strategy by introducing diagnostic substitutions at the 3′ terminus, leveraging the high sensitivity of DNA polymerase to terminal mismatches. Previous studies have demonstrated that even a single 3′-terminal mismatch can markedly inhibit polymerase extension, thereby preventing non-target amplification [36,37]. Accordingly, our terminal-mutation-driven primers yielded single, distinct amplicons for all four thrips species, with no detectable nonspecific products under mixed-template conditions. This design achieves a dynamic balance between specificity and sensitivity and demonstrates excellent reproducibility across biological replicates. Given that terminal mismatch discrimination is a universal property of DNA polymerase, this polymorphism-guided strategy has broad applicability across taxonomic groups. Because the diagnostic target is mitochondrial *COI*, which is present in both females and males, the molecular assay itself is not inherently sex-specific and is therefore applicable to both sexes in molecular identification contexts. Beyond thrips, this approach provides a scalable framework for the short-turnaround identification of other small-bodied, morphologically similar pest species and represents a conceptual shift from conserved-region-dependent barcoding to polymorphism-guided molecular diagnosis [38].

Sample preparation remains a primary bottleneck limiting the field-compatible deployment of molecular species identification workflows [39,40]. Although conventional CTAB protocols and commercial kits yield high-purity DNA, they are labor-intensive, reagent-intensive, and require specialized equipment, rendering them unsuitable for time-efficient, field-oriented workflows [41]. In contrast, the PBS-based crude extraction method demonstrated the highest success rate in direct PCR, likely attributable to its isotonic nature, buffering capacity, and high compatibility with downstream molecular reactions. This PBS-based protocol simplifies sample preparation by eliminating multi-step DNA purification and centrifugation, thereby improving compatibility with direct PCR from

 https://doi.org/10.3390/biology15020171

single insects. Standard PBS is an isotonic buffer containing Na^+, K^+, and phosphate but lacking divalent cations, providing a chemically stable environment during crude lysis. PBS can dilute potential PCR inhibitors and stabilize pH, while high-temperature incubation (98 °C) simultaneously facilitates efficient nucleic acid release and denatures endogenous proteins, including nucleases, thereby reducing nucleic acid degradation and improving amplification reproducibility [42–44]. These combined chemical and thermal effects synergistically enhance the compatibility of PBS lysates with standard DNA polymerase, enabling stable and specific amplification without purification [45]. Collectively, these mechanisms reduce reliance on conventional purification procedures, streamline sample processing, and minimize nucleic acid loss during handling [46]. Thus, PBS effectively preserves nucleic acid integrity while simplifying sample handling and enhancing assay compatibility, thereby providing reproducible templates for downstream amplification even under unpurified conditions.

Sensitivity is a key performance metric for the molecular species identification framework and is commonly quantified by the LOD, defined as the lowest target concentration that can be detected with acceptable reliability [47]. In this study, clear and reproducible amplicons were consistently obtained at a template input of 1 ng, and faint but detectable bands were observed in some thrips samples at 0.1 ng, suggesting that the LOD falls within the theoretical sensitivity range of conventional PCR, where 1–1000 ng of DNA template is typically considered workable [48]. Despite potential carryover of inhibitors from the short-turnaround lysis procedure, single, distinct amplicons were consistently obtained across all four thrips species, demonstrating that PCR inhibition was effectively alleviated under the tested dilution conditions. Although amplification remained reproducible overall, slight interspecific differences in band intensity were observed, suggesting that amplification efficiency may also be influenced by intrinsic sequence composition and structural features. GC-rich regions tend to form stable secondary structures, such as hairpins and G-quadruplexes, which hinder DNA polymerase extension under low-template conditions and thereby introduce amplification bias [49]. At limiting template levels, amplification efficiency becomes highly sensitive to primer–template binding kinetics, and even minor mismatches near the 3′ terminus can impair polymerase initiation, further amplifying interspecific variation in amplicon yield. Collectively, these molecular mechanisms plausibly account for the species-specific variation observed in amplification and indicate that the system achieves a favorable balance between sensitivity and stability.

In complex field-derived samples, the coexistence of non-target individuals and interspecific differences in metabolic profiles may introduce diverse PCR inhibitors that interfere with amplification efficiency [24,50]. To mimic this complexity, gradient mixtures of multiple thrips species were constructed to systematically evaluate the alleviation of PCR inhibition under mixed-template scenarios. Pronounced matrix-associated inhibition was observed but proved reversible. Under undiluted conditions, stable amplification was consistently observed for *F. intonsa* and *M. usitatus*, whereas *F. occidentalis* and *T. hawaiiensis* exhibited progressively weaker or undetectable bands as the proportion of non-target species increased. Following fivefold dilution, clear and distinct species-specific bands were restored for all four species, indicating that the inhibitory effect primarily resulted from excessive inhibitor concentrations in the crude lysate and confirming the pivotal role of dilution in alleviating PCR inhibition [51]. In thrips sample matrices, PCR inhibition likely arises from multiple sources acting in combination, broadly encompassing insect-derived and host plant-derived factors [24,25]. Insect-derived inhibitors include melanin and its precursors such as polyphenol oxidation products, chitin fragments, and protein residues, which can form nonspecific interactions with DNA polymerase and Mg^{2+} ions, thereby reducing enzymatic activity [24,52,53]. Additionally, components of the

 https://doi.org/10.3390/biology15020171

hemolymph phenoloxidase cascade may persist after lysis and suppress polymerase reactions [54]. This hypothesis was further supported by the single species gradient experiment, which confirmed that lysates from *F. intonsa*, *F. occidentalis*, and *M. usitatus* exerted minimal inhibitory effects on PCR, whereas *T. hawaiiensis* exhibited a non-linear amplification trend, characterized by the weakest signal at intermediate sample sizes and partial recovery at higher inputs. This pattern likely arises from the opposing effects of increasing template concentration and the accumulation of inhibitory substances in crude lysates, leading to a concentration-dependent equilibrium between amplification and suppression. After moderate dilution, the same trend persisted with overall weaker intensity, indicating that both template and inhibitor concentrations jointly determine the amplification outcome. These observations suggest that PCR inhibition in thrips lysates is influenced not only by template complexity but also by differences in biochemical composition among species, which may modulate the balance between amplification efficiency and inhibitory effects.

The alleviation of PCR inhibition observed under PBS-based extraction and moderate dilution conditions is consistent with well-established principles of PCR chemistry. Compared with lysis buffers containing chelating agents or surfactants, PBS introduces fewer exogenous compounds that may interfere with polymerase activity, allowing amplification outcomes to more directly reflect endogenous sample components [25,42]. It is well recognized that PCR inhibitors can reduce polymerase efficiency through nonspecific interactions with the enzyme or essential reaction ions such as Mg^{2+} [24,55]. Dilution reduces the effective concentration of such inhibitors and has been widely applied as a practical means to improve amplification performance in crude extracts [52,56,57]. In this context, the improved amplification observed here is in line with existing knowledge that a balance between template availability and inhibitor load determines PCR success. Together, these observations illustrate the practical utility of combining PBS-based extraction with dilution for enhancing amplification compatibility in complex insect-derived samples.

Beyond these methodological advances in DNA extraction and inhibitor management, this system has substantial practical relevance for pest thrips management. In the context of routine surveillance and large-scale monitoring, species-resolved detection supports early recognition of invasive species, real-time surveillance of population dynamics, and timely adjustment of biological control strategies and insecticide applications [58–60]. This capability is important because thrips species differ in their patterns of reproductive advantage and competitive displacement, making accurate identification before visible field symptoms appear critical for minimizing crop losses and avoiding unnecessary chemical inputs. Although adult thrips can often be identified morphologically by trained taxonomists under laboratory conditions, accurate determination becomes challenging during early infestations, in mixed-species samples, or in large-scale monitoring associated with plant trade and seedling circulation chains. Under such circumstances, the compatibility of the assay with single-adult crude lysates enables standardized and reproducible species-level identification without slide preparation or reliance on specialist taxonomic expertise, thereby facilitating routine monitoring during seedling circulation, flower transport, and trans-regional crop movement, which represent major dissemination pathways for *F. occidentalis* and other invasive thrips species and directly linking species-level diagnosis to management decisions.

Despite being reproducible under mixed-species backgrounds and strong inhibitor tolerance demonstrated across both laboratory- and field-derived samples, certain overarching limitations remain. One limitation of the present study lies in the reliance of primer design on *COI* haplotypes representing dominant regional populations, a constraint that ultimately reflects the incomplete geographic coverage and taxonomic representation of publicly available *COI* reference databases. Although additional haplotypes from geographically distant

https://doi.org/10.3390/biology15020171

regions have been reported, comprehensive incorporation of all global variants was beyond the scope of this work. This limitation is not unique to the present study but is inherent to molecular species identification frameworks based on mitochondrial *COI* markers, particularly those employing species-specific primer sets. When uncharacterized haplotypes or closely related taxa are encountered, such assays may exhibit non-amplification or, in rare cases, erroneous assignment, underscoring the importance of continued expansion and curation of reference databases. Nevertheless, within the populations examined here, the molecular species identification framework demonstrated species-specific and reproducible amplification across both single-species and mixed-species samples, supporting its practical applicability to laboratory-based surveillance workflows. Future expansion of geographic sampling and haplotype diversity will further extend the applicability of this molecular species identification framework beyond the four thrips species validated in this study [61–63]. Moreover, the assay still relies on thermocycler-based PCR and has yet to achieve true portability or real-time field-oriented workflow. Although the PBS-based protocol simplifies sample preparation, it still requires brief thermal treatment, which currently limits its full deployment under equipment-free field conditions. To overcome these constraints, emerging visual isothermal amplification technologies, such as RPA-Cas12a, have shown great promise for field-compatible molecular identification, combining minimal instrument requirements with high analytical sensitivity demonstrated across diverse insect pest and pathogen systems [64–66]. Integrating this detection framework with portable readout devices could enable a field-compatible molecular platform that automates workflows from molecular identification to pest surveillance, thereby advancing molecular identification tools toward field-adaptable and data-driven applications.

5. Conclusions

In summary, we developed a *COI*-based molecular identification framework that provides species-specific amplification with reproducible performance under mixed-species backgrounds, while reducing operational complexity and minimizing dependence on column-based DNA purification. This polymorphism-guided framework can be scalable to other minute pests and further adapted to visual isothermal amplification or CRISPR-based platforms, thereby providing a field-compatible basis for developing instrument-free, field-adaptable molecular diagnostics that support earlier and more targeted pest management decisions.

Supplementary Materials: The following supporting information can be downloaded at: https://www.mdpi.com/article/10.3390/biology15020171/s1, Figure S1: Sequence alignment of *COI* gene fragments from four thrips species; Figure S2: Primer screening using single-species DNA templates; Figure S3: Validation of primer specificity under mixed-template conditions; Figure S4: PCR amplification performance of PBS and CB-Std extraction methods at week 0 across four thrips species; Figure S5: PCR amplification performance of NCM and EDTA extraction methods at week 0 across four thrips species; Figure S6: PCR amplification performance of SDW and CB-SDS extraction methods at week 0 across four thrips species; Figure S7: PCR amplification performance of PBS and CB-Std extraction methods at week 1 across four thrips species; Figure S8: PCR amplification performance of PBS and CB-Std extraction methods at week 2 across four thrips species; Figure S9: PCR amplification performance of PBS and CB-Std extraction methods at week 3 across four thrips species; Figure S10: PCR amplification performance of PBS and CB-Std extraction methods at week 4 across four thrips species; Figure S11: Temporal variation of A260/A280 absorbance ratios across extraction methods; Figure S12: Temporal variation of A260/A230 absorbance ratios across extraction methods; Table S1: List of primers employed in molecular characterization.

Author Contributions: Conceptualization, Q.Q., J.C., Y.C., H.W. (Hui Wei) and W.H.; methodology, Q.Q., J.C., Y.C., T.C., H.F., Y.M.S., H.W. (Han Wang) and L.T.; software, Q.Q. and J.C.; validation, Q.Q.,

 https://doi.org/10.3390/biology15020171

J.C., Y.C., T.C. and H.F.; formal analysis, Q.Q., J.C. and Y.C.; investigation, Q.Q., J.C., Y.C., T.C., H.F., Y.M.S., H.W. (Han Wang), L.T., H.Z. and Z.C.; resources, H.W. (Hui Wei), W.H. and T.L.; data curation, Q.Q., J.C., H.F. and Y.M.S.; writing—original draft preparation, Q.Q.; writing—review and editing, J.C., Y.C., H.W. (Hui Wei) and W.H.; visualization, Q.Q. and J.C.; supervision, H.W. (Hui Wei), W.H. and T.L.; project administration, H.W. (Hui Wei) and W.H.; funding acquisition, H.W. (Hui Wei) and W.H. All authors have read and agreed to the published version of the manuscript.

Funding: This work was supported by the Major Project of Science and Technology of Fujian Province (2024NZ029029) and the Fujian Provincial Science and Technology Project (2023L3021 and 2025N0081).

Institutional Review Board Statement: Not applicable.

Informed Consent Statement: Not applicable.

Data Availability Statement: The data presented in this study are available upon reasonable request from the corresponding author.

Conflicts of Interest: The authors declare no conflicts of interest.

References

1. Kumar, B.; Omkar. Thrips. In *Polyphagous Pests of Crops*; Omkar, Ed.; Springer: Singapore, 2021; pp. 373–407. [CrossRef]
2. Morse, J.G.; Hoddle, M.S. Invasion biology of thrips. *Annu. Rev. Entomol.* **2006**, *51*, 67–89. [CrossRef] [PubMed]
3. Rodríguez, D.; Coy-Barrera, E. Overview of updated control tactics for western flower thrips. *Insects* **2023**, *14*, 649. [CrossRef] [PubMed]
4. Manideep, S.; Muthuswami, M.; Shanmugam, P.S.; Tulasi, B.; Kumar, T. Invasive thrips, *Thrips parvispinus* (Karny, 1992), and its threat to agriculture: A comprehensive review. *Plant Arch.* **2025**, *25*, 1118–1127. [CrossRef]
5. He, Z.; Guo, J.; Reitz, S.R.; Lei, Z.; Wu, S. A global invasion by the thrip, *Frankliniella occidentalis*: Current virus vector status and its management. *Insect Sci.* **2020**, *27*, 626–645. [CrossRef]
6. Zhao, K.; Rosa, C. Thrips as the transmission bottleneck for mixed infection of two orthotospoviruses. *Plants* **2020**, *9*, 509. [CrossRef]
7. Kumar, V.; Kakkar, G.; Seal, D.R.; McKenzie, C.L.; Colee, J.; Osborne, L.S. Temporal and spatial distribution of an invasive thrips species *Scirtothrips dorsalis* (Thysanoptera: Thripidae). *Crop Prot.* **2014**, *55*, 80–90. [CrossRef]
8. Buitenhuis, R.; Jandricic, S.; Scott-Dupree, C.; Summerfield, A.; Labbé, R. *Frankliniella occidentalis* (Pergande), western flower thrips/*Thrips des petits fruits*, *Thrips tabaci* (Lindeman), onion thrips/*Thrips de l'oignon* (Thysanoptera: Thripidae), and other pest thrips in greenhouse crops. In *Biological Control Programmes in Canada, 2013–2023*; CABI: Wallingford, UK, 2024; pp. 232–242. [CrossRef]
9. Tomizawa, Y.; Aizawa, M.; Jouraku, A.; Sonoda, S. Field survey of reproductive modes and sodium channel mutations associated with pyrethroid resistance in *Thrips tabaci*. *J. Pestic. Sci.* **2024**, *49*, 122–129. [CrossRef]
10. Mirab-balou, M. Thysanoptera (Insecta) of China: An updated checklist. *J. Insect Biodivers. Syst.* **2025**, *11*, 469–541. [CrossRef]
11. Ntirenganya, E.; Xie, Y.; Mao, J.; Hussain, M.; Zhang, H. Species richness and seasonal abundance of thrips (Thysanoptera) on different tea cultivars in Xishuangbanna, Yunnan Province, China. *Int. J. Entomol. Res.* **2020**, *5*, 197–203.
12. Fu, B.; Tao, M.; Xue, H.; Jin, H.; Liu, K.; Qiu, H.; Yang, S.; Yang, X.; Gui, L.; Zhang, Y.; et al. Spinetoram resistance drives interspecific competition between *Megalurothrips usitatus* and *Frankliniella intonsa*. *Pest Manag. Sci.* **2022**, *78*, 2129–2140. [CrossRef]
13. Tang, L.; Guo, L.; Wu, J.; Zang, L. Thrips in genus *Megalurothrips* (Thysanoptera: Thripidae): Biodiversity, bioecology, and IPM. *J. Integr. Pest Manag.* **2023**, *14*, 8. [CrossRef]
14. Yang, C.; Qiao, F.; Lu, Z.; Li, C.; Liu, T.; Gao, Y.; Zhang, B. Interspecific competitions between *Frankliniella intonsa* and *Frankliniella occidentalis* on fresh lentil bean pods and pepper plants. *Insects* **2023**, *14*, 424. [CrossRef] [PubMed]
15. Li, X.; Geng, S.; Zhang, Z.; Zhang, J.; Li, W.; Huang, J.; Lin, W.; Bei, Y.; Lu, Y. Species-specific aggregation pheromones contribute to coexistence in two closely related thrips species. *Bull. Entomol. Res.* **2019**, *109*, 119–126. [CrossRef] [PubMed]
16. Zhao, X.; Reitz, S.R.; Yuan, H.; Lei, Z.; Paini, D.R.; Gao, Y. Pesticide-mediated interspecific competition between local and invasive thrips pests. *Sci. Rep.* **2017**, *7*, 40512. [CrossRef]
17. Ghosh, A.; Jangra, S.; Dietzgen, R.G.; Yeh, W. Frontiers approaches to the diagnosis of thrips (Thysanoptera): How effective are the molecular and electronic detection platforms? *Insects* **2021**, *12*, 1014. [CrossRef]
18. Mehle, N.; Trdan, S. Traditional and modern methods for the identification of thrips (*Thysanoptera*) species. *J. Pest Sci.* **2012**, *85*, 179–190. [CrossRef]

 https://doi.org/10.3390/biology15020171

19. Cao, Y.; Zhi, J.; Zhang, R.; Li, C.; Liu, Y.; Lv, Z.; Gao, Y. Different population performances of *Frankliniella occidentalis* and *Thrips hawaiiensis* on flowers of two horticultural plants. *J. Pest Sci.* **2018**, *91*, 79–91. [CrossRef]
20. Smith-Pardo, A.H.; Pérez, L.M.; Benítez, H.A. Unlocking species identity: Geometric morphometrics of head and thorax shapes in invasive and non-invasive quarantine-significant thrips (Thysanoptera: Terebrantia). *Front. Insect Sci.* **2025**, *5*, 1558242. [CrossRef]
21. Khatun, M.F.; Hwang, H.; Kang, J.; Lee, K.; Kil, E. Genetic diversity and DNA barcoding of thrips in Bangladesh. *Insects* **2024**, *15*, 107. [CrossRef]
22. Rugman-Jones, P.F.; Hoddle, M.S.; Stouthamer, R. Nuclear-mitochondrial barcoding exposes the global pest western flower thrips (Thysanoptera: Thripidae) as two sympatric cryptic species in its native California. *J. Econ. Entomol.* **2010**, *103*, 877–886. [CrossRef]
23. Jangra, S.; Ghosh, A.; Mukherjee, S.; Baranwal, V.K.; Dietzgen, R.G. Development of a polymerase spiral reaction-based isothermal assay for rapid identification of *Thrips palmi*. *Front. Mol. Biosci.* **2022**, *9*, 853339. [CrossRef]
24. Schrader, C.; Schielke, A.; Ellerbroek, L.; Johne, R. PCR inhibitors—Occurrence, properties and removal. *J. Appl. Microbiol.* **2012**, *113*, 1014–1026. [CrossRef]
25. Abdel-Latif, A.; Osman, G. Comparison of three genomic DNA extraction methods to obtain high DNA quality from maize. *Plant Methods* **2017**, *13*, 1. [CrossRef]
26. Lin, T.; You, Y.; Zeng, Z.; Chen, Y.; Hu, J.; Lin, S.; Hu, Q.; Yang, F.; Wei, H. Temperature-dependent demography of *Thrips hawaiiensis* (Thysanoptera: Thripidae): Implications for prevention and control. *Environ. Entomol.* **2021**, *50*, 1455–1465. [CrossRef] [PubMed]
27. Li, D.; Sooda, A.; Gunawardana, D.; Thomas, A.; Chen, Y.; Kumarasinghe, L. DNA barcodes for thrips species and development of multiplex real-time PCR assay for *Frankliniella occidentalis* Pergande, *Frankliniella panamensis* Hood, *Thrips palmi* Karny and *Thrips tabaci* Lindeman (Thysanoptera: Thripidae). *N. Z. Entomol.* **2023**, *46*, 16–34. [CrossRef]
28. Musa, S.; Ladányi, M.; Loredo Varela, R.C.; Fail, J. A morphometric analysis of the *Thrips tabaci Lindeman* species complex (Thysanoptera: Thripidae). *Arthropod Struct. Dev.* **2023**, *72*, 101228. [CrossRef] [PubMed]
29. Sharma, M.; Sridhar, J.; Akanand, D.; Sivalingam, P.N.; Bhowmick, A.K. Morphological and molecular characterization of cryptic and invasive thrips species in Central India. *Int. J. Trop. Insect Sci.* **2024**, *44*, 2305–2314. [CrossRef]
30. Crespi, B.J.; Carmean, D.A.; Mound, L.A.; Worobey, M.; Morris, D. Phylogenetics of social behavior in Australian gall-forming thrips: Evidence from mitochondrial DNA sequence, adult morphology and behavior, and gall morphology. *Mol. Phylogenet. Evol.* **1998**, *9*, 163–180. [CrossRef]
31. Chao, H.; Dietzgen, R.G.; Thomas, J.E.; Geering, A.D.W. New molecular diagnostic and surveillance tools for orthotospoviruses and their thrips vectors. *Phytopathol. Res.* **2025**, *7*, 45. [CrossRef]
32. Lamichhane, B.; Brockway, C.; Evasco, K.; Nicholson, J.; Neville, P.J.; Mackenzie, J.S.; Smith, D.; Imrie, A. DNA barcoding for the identification of adult mosquitoes (Diptera: Culicidae) in Western Australia. *Ecol. Evol.* **2024**, *14*, e70493. [CrossRef]
33. Chen, H.; Dong, H.; Yuan, H.; Shan, W.; Zhou, Q.; Li, X.; Peng, H.; Ma, Y. Mitochondrial *COI* and *Cytb* gene as valid molecular identification marker of sandfly species (Diptera: Psychodidae) in China. *Acta Trop.* **2023**, *238*, 106798. [CrossRef]
34. Elbrecht, V.; Braukmann, T.W.A.; Ivanova, N.V.; Prosser, S.W.J.; Hajibabaei, M.; Wright, M.; Zakharov, E.V.; Hebert, P.D.N.; Steinke, D. Validation of *COI* metabarcoding primers for terrestrial arthropods. *PeerJ* **2019**, *7*, e7745. [CrossRef]
35. Penel, B.; Meynard, C.N.; Benoit, L.; Bourdonné, A.; Clamens, A.; Soldati, L.; Migeon, A.; Chapuis, M.; Piry, S.; Kergoat, G.; et al. The best of two worlds: Toward large-scale monitoring of biodiversity combining *COI* metabarcoding and optimized parataxonomic validation. *Ecography* **2025**, *2025*, e07699. [CrossRef]
36. Huang, K.; Zhang, J.; Li, J.; Qiu, H.; Wei, L.; Yang, Y.; Wang, C. Exploring the impact of primer–template mismatches on PCR performance of DNA polymerases varying in proofreading activity. *Genes* **2024**, *15*, 215. [CrossRef]
37. Kalendar, R.; Shustov, A.V.; Akhmetollayev, I.; Kairov, U. Designing allele-specific competitive-extension PCR-based assays for high-throughput genotyping and gene characterization. *Front. Mol. Biosci.* **2022**, *9*, 773956. [CrossRef]
38. Xu, G.; Yang, H.; Qiu, J.; Reboud, J.; Zhen, L.; Ren, W.; Xu, H.; Cooper, J.M.; Gu, H. Sequence terminus-dependent PCR for site-specific mutation and modification detection. *Nat. Commun.* **2023**, *14*, 1169. [CrossRef] [PubMed]
39. Ingle, R.G.; Zeng, S.; Jiang, H.; Fang, W. Current developments of bioanalytical sample preparation techniques in pharmaceuticals. *J. Pharm. Anal.* **2022**, *12*, 517–529. [CrossRef] [PubMed]
40. Nichols, Z.E.; Geddes, C.D. Sample Preparation and Diagnostic Methods for a Variety of Settings: A Comprehensive Review. *Molecules* **2021**, *26*, 5666. [CrossRef]
41. Liu, B.; Wu, H.; Cao, Y.; Ma, G.; Zheng, X.; Zhu, H.; Song, X.; Sui, S. Reducing costs and shortening the cetyltrimethylammonium bromide (CTAB) method to improve DNA extraction efficiency from wintersweet and some other plants. *Sci. Rep.* **2025**, *15*, 13441. [CrossRef] [PubMed]
42. Thongjued, K.; Chotigeat, W.; Bumrungsri, S.; Thanakiatkrai, P.; Kitpipit, T. A new cost-effective and fast direct PCR protocol for insects based on PBS buffer. *Mol. Ecol. Resour.* **2019**, *19*, 691–701. [CrossRef]
43. Kitpipit, T.; Chotigeat, W.; Linacre, A.; Thanakiatkrai, P. Forensic animal DNA analysis using economical two-step direct PCR. *Forensic Sci. Med. Pathol.* **2014**, *10*, 29–38. [CrossRef]

 https://doi.org/10.3390/biology15020171

44. Loto, F.; Romero, C.M.; Baigorí, M.D.; Pera, L.M. Direct DNA amplification from fall armyworm (Lepidoptera: Noctuidae) samples. *Fla. Entomol.* **2013**, *96*, 1618–1620. [CrossRef]
45. Zhang, Z.; Kermekchiev, M.B.; Barnes, W.M. Direct DNA amplification from crude clinical samples using a PCR enhancer cocktail and novel mutants of Taq. *J. Mol. Diagn.* **2010**, *12*, 152–161. [CrossRef] [PubMed]
46. Aslam, M.; Naeem, F.; Seher, R.; Shabbir, M.Z.; Shehzad, W.; Imran, M. Effect of storage temperature and duration on direct PCR amplification of various feather types and DBS matrices. *Gene* **2023**, *854*, 147116. [CrossRef]
47. Foo, P.C.; Najian, A.B.N.; Muhamad, N.A.; Ahamad, M.; Mohamed, M.; Yean Yean, C.; Lim, B.H. Loop-mediated isothermal amplification (LAMP) reaction as viable PCR substitute for diagnostic applications: A comparative analysis study of LAMP, conventional PCR, nested PCR (nPCR) and real-time PCR (qPCR) based on *Entamoeba histolytica* DNA derived from faecal sample. *BMC Biotechnol.* **2020**, *20*, 34. [CrossRef]
48. Hunter, M.E.; Dorazio, R.M.; Butterfield, J.S.S.; Meigs-Friend, G.; Nico, L.G.; Ferrante, J.A. Detection limits of quantitative and digital PCR assays and their influence in presence–absence surveys of environmental DNA. *Mol. Ecol. Resour.* **2017**, *17*, 221–229. [CrossRef] [PubMed]
49. Stevens, A.J.; Stuffrein-Roberts, S.; Cree, S.L.; Gibb, A.; Miller, A.L.; Doudney, K.; Aitchison, A.; Eccles, M.R.; Joyce, P.R.; Filichev, V.V.; et al. G-Quadruplex structures and CpG methylation cause drop-out of the maternal allele in polymerase chain reaction amplification of the imprinted *MEST* gene promoter. *PLoS ONE* **2014**, *9*, e113955. [CrossRef]
50. Rossen, L.; Nørskov, P.; Holmstrøm, K.; Rasmussen, O.F. Inhibition of PCR by components of food samples, microbial diagnostic assays and DNA-extraction solutions. *Int. J. Food Microbiol.* **1992**, *17*, 37–45. [CrossRef]
51. McKee, A.M.; Spear, S.F.; Pierson, T.W. The effect of dilution and the use of a post-extraction nucleic acid purification column on the accuracy, precision, and inhibition of environmental DNA samples. *Biol. Conserv.* **2015**, *183*, 70–76. [CrossRef]
52. Opel, K.L.; Chung, D.; McCord, B.R. A study of PCR inhibition mechanisms using real-time PCR. *J. Forensic Sci.* **2010**, *55*, 25–33. [CrossRef]
53. Vajpayee, K.; Srivastava, S.; Sharma, S.; Gupta, S.; Srivastava, A.; Paida, V.; Dash, H.; Pappachan, A.; Shukla, R. Mechanistic insights into melanin-induced PCR inhibition and its NanoPCR-based mitigation. *Research Square* **2025**. Preprint. [CrossRef]
54. Lu, Z.; Jiang, H. Regulation of phenoloxidase activity by high- and low-molecular-weight inhibitors from the larval hemolymph of *Manduca sexta*. *Insect Biochem. Mol. Biol.* **2007**, *37*, 478–485. [CrossRef] [PubMed]
55. Wang, J.; Konigsberg, W.H. Two-metal-ion catalysis: Inhibition of DNA polymerase activity by a third divalent metal ion. *Front. Mol. Biosci.* **2022**, *9*, 824794. [CrossRef] [PubMed]
56. Zafeiriadou, A.; Nano, K.; Thomaidis, N.S.; Markou, A. Evaluation of PCR-enhancing approaches to reduce inhibition in wastewater samples and enhance viral load measurements. *Sci. Total Environ.* **2024**, *955*, 176768. [CrossRef]
57. Hamza, I.A.; Leifels, M. Assessment of PCR inhibitor removal methods to monitor viruses in environmental water samples: DAX-8 outperforms competitors. *Water Air Soil Pollut.* **2023**, *235*, 20. [CrossRef]
58. Jaybhaye, S.G.; Chavhan, R.L.; Hinge, V.R.; Deshmukh, A.S.; Kadam, U.S. CRISPR-Cas assisted diagnostics of plant viruses and challenges. *Virology* **2024**, *597*, 110160. [CrossRef]
59. Lv, C.; Zhang, F.; Ren, L.; Zhu, P.; Cheng, X.; Yang, X.; Chen, C.; Liu, B. Rapid visual detection assay for *Bactrocera dorsalis* (Hendel) using recombinase polymerase amplification and CRISPR/Cas12b. *Sci. Rep.* **2025**, *15*, 17328. [CrossRef]
60. Alon, D.M.; Partosh, T.; Burstein, D.; Pines, G. Rapid and sensitive on-site genetic diagnostics of pest fruit flies using CRISPR-Cas12a. *Pest Manag. Sci.* **2023**, *79*, 68–75. [CrossRef] [PubMed]
61. Gaytán, Á.; Bergsten, J.; Canelo, T.; Pérez-Izquierdo, C.; Santoro, M.; Bonal, R. DNA barcoding and geographical scale effect: The problems of undersampling genetic diversity hotspots. *Ecol. Evol.* **2020**, *10*, 10754–10772. [CrossRef]
62. Šamulková, M.; Čiamporová-Zaťovičová, Z.; Čiampor, F.; Tuhrinová, K.; Macko, P. Is the DNA barcode database fit for purpose? Assessing the feasibility of reverse taxonomy in five key aquatic insect orders relevant to freshwater biomonitoring under the Water Framework Directive. *Biologia* **2025**, *80*, 2733–2742. [CrossRef]
63. Lo, E.; Nie, R.E.; Vogler, A.P. The geographic and phylogenetic structure of public DNA barcode databases: An assessment using *Chrysomelidae* (leaf beetles). *Front. Ecol. Evol.* **2024**, *12*, 1305898. [CrossRef]
64. Ji, T.; Fang, X.; Gao, Y.; Yu, K.; Gao, X. Research progress on the application of RPA-CRISPR/Cas12a in the rapid visual detection of pathogenic microorganisms. *Front. Cell Infect. Microbiol.* **2025**, *15*, 1640938. [CrossRef] [PubMed]
65. Shashank, P.R.; Parker, B.M.; Rananaware, S.R.; Plotkin, D.; Couch, C.; Yang, L.G.; Nguyen, L.T.; Prasannakumar, N.R.; Braswell, W.E.; Jain, P.K.; et al. CRISPR-based diagnostics detects invasive insect pests. *Mol. Ecol. Resour.* **2024**, *24*, e13881. [CrossRef]
66. Chen, Y.; Shi, M.; Chen, Y.; Zhao, J.; Yang, X.; Fu, J.; Desneux, N.; Li, J. Rapid and equipment-free identification of papaya mealybug *Paracoccus marginatus* based on RPA-CRISPR/Cas12a. *Pest Manag. Sci.* **2025**, *81*, 230–239. [CrossRef] [PubMed]

 https://doi.org/10.3390/biology15020171

biology

MDPI

Article

Insecticidal Activity of *Angelica archangelica* Essential Oil and Transcriptomic Analysis of *Sitophilus zeamais* in Response to Oil Fumigation

Guochang Wang [1,†], Xing Ge [1,*,†], Dongbiao Lü [1,*], Ziyue Zhang [1], Li Wang [1], Saili Sun [1], Xiaoyi Jia [1], Baizhong Zhang [1,2], Kassen Kuanysh [3], Sarsekova Dani [4] and Hongliang Wang [1]

1 Henan Province Engineering Research Center of Biological Pesticide & Fertilizer Development and Synergistic Application, College of Plant Protection and Environment, Henan Institute of Science and Technology, Xinxiang 453003, China; wgchslbh@163.com (G.W.); zzy_94983@163.com (Z.Z.); 15617179499@163.com (L.W.); s17839425509@163.com (S.S.); jxy2025hist@163.com (X.J.); baizhongok@163.com (B.Z.); wanghlzb@163.com (H.W.)
2 Hebi Institute of Engineering and Technology, Henan Polytechnic University, Hebi 458030, China
3 Plant Biotechnology Center, Kazakh National Agrarian Research University, Almaty 050040, Kazakhstan; kassenkuanysh12@gmail.com
4 Faculty of Forestry and Land Resources, Kazakh National Agrarian Research University, Almaty 050040, Kazakhstan; dani999@mail.ru
* Correspondence: ge_xing_22@126.com (X.G.); dongbiao_0613@163.com (D.L.)
† These authors contributed equally to this work.

Simple Summary: The maize weevil, *Sitophilus zeamais*, is a globally distributed stored-grain pest that causes significant losses to cereals, posing a serious threat to food security. We investigated chemical composition and fumigation activity of *Angelica archangelica* root essential oil (AAEO) against *S. zeamais*. The main components of AAEO and its median lethal concentration (LC_{50}) for fumigation against this pest were identified. The enzymatic activities of acetylcholinesterase, glutathione-S-transferase, and carboxylesterase were measured when fumigated by AAEO LC_{50}. The study demonstrates that AAEO exhibits strong insecticidal activity against *S. zeamais*, affecting survival, detoxification enzymes, and gene transcription. The present study provide a scientific basis for the development of effective plant essential oil-based insecticide against *S. zeamais*.

Abstract: *Sitophilus zeamais* is one of the most destructive pests of stored grains. Both adults and larvae penetrate and consume the grains, thereby diminishing the grain quality and nutritional value. We determined the chemical composition of *Angelica archangelica* essential oil, its fumigation toxicity against *S. zeamais*, and its effects on the activities of detoxification enzymes in the insects. RNA-seq was performed to analyze the impact of the essential oil on the transcriptional level of *S. zeamais*, and qRT-PCR was conducted to validate the differentially expressed genes. Chemical analysis identified 35 components in essential oil, including δ-3-Carene (24.26%), Limonene (19.81%), and α-Pinene (14.96%). A significant positive correlation was observed between the fumigation activity of the essential oil and the applied dose. The median lethal concentrations (LC_{50}) values were 164.38, 132.62, and 90.35 mg/L air at 24, 48, and 72 h, respectively. Fumigation significantly inhibited the activities of the three detoxification enzymes. RNA-seq revealed a total of 3718 significantly differentially expressed genes (DEGs). qRT-PCR confirmed that the expression patterns of the DEGs were consistent with the RNA-seq data. This study comprehensively evaluates the control efficacy of *A. archangelica* essential oil against *S. zeamais* and provides data support for developing novel, eco-friendly, plant-based pesticides.

Keywords: *Sitophilus zeamais*; essential oil; insecticidal activity; enzymatic activity; RNA-seq

 https://doi.org/10.3390/biology14111606

1. Introduction

Sitophilus zeamais is a destructive pest of stored grain, including rice, maize, wheat and their products, causing extensive loss due to reduction in quality and nutritional value [1,2]. Adults of *S. zeamais* can directly lay eggs inside the grains to protect the developing larvae, resulting in grain losses and contamination. The larvae infest the grains, causing entrance of various fungi and bacteria, thereby increasing the risk of contamination [3]. To combat postharvest losses, chemical fumigants including methyl bromide and phosphine were frequently and widely used to control *S. zeamais* during storage [4,5]. Although they have brought remarkable economic benefits, long-term use of these chemicals has led to the development of pest resistance. Additionally, the chemical fumigants, such as methyl bromide, have been prohibited due to their environmental pollution [6,7]. Thus, it is critical to explore natural products as potential bioinsecticide sources to control stored-product pests.

Plant essential oil (EO), as one class of important volatile secondary metabolites, is extracted from flowers, seeds, leaves, roots and fruits of plants [8]. EOs exhibits not only high bioactivity but also outstanding advantages such as remarkable efficacy, low toxicity, and short persistence in ecosystems [9]. Up to now, many plant essential oils have been reported to have insecticidal activity [10–12]; the effects on *S. zeamais* could be classified as repellency, fumigation, contact action. Therefore, EOs can be regarded as a foundation of the development of biopesticides [13]. Plant essential oils, as an emerging "green insecticide," demonstrate significant potential in the field of stored-grain pest control [9]. Research has progressed from simple toxicity assays to in-depth investigations of complex multi-target mechanisms, providing a compelling scientific foundation for their use as alternatives to traditional chemical fumigants [2]. In comparison, their key advantages include high efficacy and broad-spectrum activity, multi-targeted mechanisms of action, environmental friendliness and biodegradability, relative safety for non-target organisms, and a low tendency to induce pest resistance [7]. Comprehensive research positions them as a promising future direction for integrated pest management.

Angelica archangelica L. (syn. *A. officinalis* Hoffm.), well known as garden angelica and folk medicine, is a bi-annual or perennial herb with a strong aroma, so it is often used as a food ingredient. The essential oil from roots of *A. archangelica* has multiple biological activities, including antioxidant and antibacterial effects, among other pharmacological effects [14–16]. However, there is a lack of evidence regarding the insecticidal activity of *A. archangelica* root essential oil against *S. zeamais*.

This study investigates the fumigant toxicity of *A. archangelica* root essential oil against *S. zeamais* adults and evaluates its impact on the insect's detoxification and nervous system enzymes. Furthermore, we performed transcriptome sequencing on *S. zeamais* after fumigation with essential oil for 24 h to identify the genes altered by treatment. The results provide a theoretical basis for the management of stored-grain pests using *A. archangelica* root essential oil.

2. Materials and Methods

2.1. Essential Oil

The essential oil from *A. archangelica* was obtained from Ji'An HuaTianBao Herbs Biological Products Factory (purity $\geq$ 95.00%; Ji'an, China). It was extracted from root of *A. archangelica* by hydro-distillation.

2.2. Gas Chromatography and Mass Spectrometry (GC-MS)

Components of the essential oil of *A. archangelica* were identified by a GC-MS 7890B-5977B system (Agilent Technologies, Palo Alto, CA, USA) coated with a DB-5 MS fused silica capillary column (30 m × 0.25 mm × 0.25 μm). n-Hexane was used as the dilution solvent. The injection temperature was set at 250 °C, with 1 μL of a 10% solution injected each time. The temperature was maintained at 50 °C for 2 min, then increased at a rate of 10 °C per minute to 240 °C, and held for 5 min. The carrier gas was helium at flow rate of 1.0 μL/min. Spectra were scanned from 50 to 550 m/z. The retention indices were determined in relation to a homologous series of n-alkanes (C7–C40) under the same operating conditions. The components were identified by NIST 14.L and confirmed by comparing the Arithmetic Indices and comparison with authentic standards. Relative percentages of the individual components of the essential oil were quantified on the basis of the peak area.

2.3. Insect Culture

S. zeamais were maintained in the insectarium at Henan Institute of Science and Technology for over three years without exposure to insecticides. The insects were reared in a round glass jar with a 10 cm diameter and a height of 15 cm, containing 0.5 kg of sterilized whole wheat. The glass jars were placed in an incubator maintained at 28 ± 1 °C and 75 ± 5% RH, in total darkness.

2.4. Fumigant Toxicity

S. zeamais adults were used to evaluate the fumigant activity of *A. archangelica* essential oils (EOs). Solutions of *A. archangelica* EOs were prepared at concentrations of 34, 68, 102, 136, and 170 mg/L using hexane. A 100 μL aliquot of each concentration was dropped onto a Whatman filter paper strip (1.5 cm × 5.0 cm) and evaporated for 1 min at room temperature. It was then suspended in a triangular flask (50 mL). Thirty randomly selected *S. zeamais* adults were released into a fumigation bottle containing 3 g of wheat, and the cap was screwed tightly shut. The control consisted of hexane solvent. The numbers of dead *S. zeamais* adults were observed and recorded at 24, 48, and 72 h post-treatment. All treatments and controls were conducted independently five times.

2.5. Assessment of Enzyme Activity

We determined the activities of acetylcholinesterase (AChE), glutathione S-transferase (GST), and carboxylesterase (CarE). A series of treatments to *S. zeamais* were set with dose of 34, 68, 102, 136, and 170 mg/L of *A. archangelica* EOs, respectively. Living insects were collected at 24 h after EO treatment, and used for the extraction of total protein. In order to explore the enzyme activity in different stages, another set of test samples were treated with LC_{50} of EO (164.38 mg/L) at 4, 8, 12, 24, and 48 h, respectively. The live insects after treatment were weighed, rinsed, and homogenized, followed by the addition of 900 μL of pre-cooled normal saline. The mixture was centrifuged at 2500 rpm for 10 min at 4 °C. The supernatant was stored at −80 °C. The entire experimental procedure was performed on ice. The concentration of total protein was determined using the total protein quantitative assay (Nanjing Jiancheng Bioengineering Institute, Nanjing, China). The activity of each enzyme, including AChE, GST, and CarE was tested following the instruction of the AChE, GST, and CarE assay kits (Nanjing Jiancheng Bioengineering Institute), respectively. Three replicates were performed for each treatment and each replicate was performed three times.

2.6. RNA Extraction and Sequencing

In order to explore the functional genes of *S. zeamais* against *A. archangelica* EOs, 164.83 mg/L (LC_{50} of 24 h) EOs was used to fumigate the adults for 24 h. The control was hexane solvent. Living insects were collected and frozen separately in liquid nitrogen and stored at −80 °C for RNA extraction.

Total RNA was sequentially extracted using Trizol reagent (Invitrogen, Carlsbad, CA, USA), followed by RNA integrity assessment with the Bioanalyzer 2100 (Agilent Technologies, Santa Clara, CA, USA), and subsequent library construction.

2.7. Transcript Assembly and Gene Function Annotation

Raw reads were firstly processed through removing low-quality reads, reads containing adapter, and reads containing N base to obtain clean reads. Transcriptome assembly was accomplished using Trinity with min_kmer_cov set to 2 by default and all other parameters set default [17]. To obtain the comprehensive information of gene functions, assembled unigenes were annotated according to Nr, Nt, Pfam, KOG/COG, Swiss-Prot, KO, and GO database.

2.8. Differentially Expressed Genes (DEGS) Analysis

To identify DEGs in *A. archangelica* EOs fumigated *S. zeamais*, clean reads were analyzed by DESeq2 R package (1.20.0) to calculate the Padj of differential expression [18]. In addition, in order to control error detection rate (FDR), a normalized absolute log2-fold change (fumigated/control) of 2 with adjusted p-value less than 0.05 were set as the threshold for significantly differentially expression.

2.9. GO and KEGG Pathway Enrichment Analysis

After classifying the annotation information, the DEGs were associated with the Gene Ontology (GO) for enrichment analysis. GO functional enrichment was considered as the differential expression of genes with a significant enrichment threshold of padj less than 0.05. Kyoto Encyclopedia of Genes and Genomes (KEGG) is a comprehensive database integrating genomic, chemical, and system functional information. The threshold for KEGG pathway enrichment was padj less than 0.05. The ClusterProfiler (3.8.1) software was used to conduct GO functional enrichment analysis and KEGG pathway enrichment analysis on the DEGs [19].

2.10. Real Time Quantitative Reverse Transcription PCR (qRT-PCR) Analysis

From the DEGs, 14 genes were selected and a heat map was generated using the heatmap package in R software (Version 3.5.0) based on their expression levels in the treatment and control groups, including five P450 genes, three UGT genes, three GST genes, two CarE genes, and one JHEH gene. qRT-PCR was performed on a Real-time Detection System (QuantStuido 5; Waltham, USA) to verify the reliability of transcriptome. qRT-PCR was performed using a 20 μL reaction volume. The program conditions were 95 °C for 30 s, 40 cycles at 95 °C for 10 s, and 60 °C for 30 s. The primers used are listed in Table 1. Glyceraldehyde 3-phosphate dehydrogenase (GAPDH) was used as a reference gene [20]. Three technical repeats were performed for each sample. The relative expression level of genes was performed using $2^{-\Delta\Delta CT}$.

Table 1. qRT-PCR primers and internal reference gene sequences.

Primer Name	Primer Sequence (5′-3′)
SzeaCYP6A14	F: TCGCTGCTGGATATGAGA R: TGGATTGAACTTGTCAGGAT
SzeaCYP6M5	F: CTTAGTATCAGCCATCCTAGT R: TTGAGTTGCCTGTATTGTTC
SzeaCYP6A19	F: GCTAGGCGAGGAATCTTC R: CGTAAGCAACAGCAGAATC
SzeaCYP12V2	F: GGCTCAGAACAACTCAAGA R: GTCCAACCTCCAATTCCAA
SzeaCYP6A13	F: ACGAGAAGCGACAGGATT R: GAAGAGTGGTTGCGGATG
SzeaUGT2C1	F: GTCTGTATCCTGCTCATCC R: TGTCCAAGAACTCCTGTAAG
SzeaUGT2C10	F: ACCACGACAAGAACTTCA R: TGGAACAGTCTGGACCTT
SzeaUGT28C1	F: GCACCTTCCATAGTGATGA R: CCACGCATTAACAATTCTCT
SzeaGST1	F: TGTACGGTATTAAGGCTAGTC R: CAGGAATGGTATGTTGAGGA
SzeaGST2	F: TAATCTGTCGGAGAAGGAAC R: TCGGATAAAGGTCGTCATTT
SzeaGST3	F: GCTCTGGAACTTCTCAACA R: TTCACCTGCGGATAGTCA
SzeaCarE1	F: CGAATCTGTTGAGCAAGG R: CAGTTGAGCCGTTGTAAG
SzeaCarE2	F: AAGGAGAGTTAGGAGAATCTG R: CGTGTTAGTTCATCGTCATC
SzeaJHEH1	F: GATGGTGGACTGACAAGA R: GACTACAGCCAGAAGGAAT
GAPDH	F: AACTTTGCCGACAGCCTTGG R: GCGCCCATGTATGTAGTTGG

2.11. Statistical Analysis

The percentage of insect mortality was transformed using the arcsine square root transformation. All experimental data were analyzed using SPSS Statistics 22.0 (https://www.ibm.com/products/spss-statistics) with this method. Based on the assumptions of normality and homogeneity of variance, one-way analysis of variance (ANOVA) was subsequently performed, followed by Tukey's honestly significant difference (HSD) test. Differences were considered statistically significant at $p < 0.05$.

3. Results

3.1. Chemical Composition of Essential Oil

The chemical composition of the essential oil is summarized in Table 2. A total of 35 compositions accounting for 96.94% of *A. archangelica* essential oil were identified. The major components were δ-3-Carene (24.26%), Limonene (19.81%) and α-Pinene (14.96%), followed by o-Cymene (9.94%), β-Pinene (6.55%), Linalool (6.49%), Myrcene (5.60%).

Table 2. Chemical composition of *A. archangelica* essential oil.

Peak NO.	AI ^{a}	Compounds	%RA ^{b}	Identification Method ^{c}
1	918	Tricyclene	0.15	MS, AI
2	922	Artemisia triene	0.13	MS, AI
3	932	α-Pinene	14.96	MS, AI, AS
4	947	Camphene	1.50	MS, AI

Table 2. *Cont.*

Peak NO.	AI [a]	Compounds	%RA [b]	Identification Method [c]
5	967	Sabinene	0.48	MS, AI
6	973	β-Pinene	6.55	MS, AI
7	979	*trans*-Isolimonene	0.23	MS, AI
8	988	Myrcene	5.60	MS, AI, AS
9	998	δ-2-Carene	0.38	MS, AI
10	1010	δ-3-Carene	24.26	MS, AI
11	1013	1,4-Cineole	1.38	MS, AI
12	1023	o-Cymene	9.94	MS, AI
13	1031	Limonene	19.81	MS, AI, AS
14	1068	n-Octanol	0.12	MS, AI
15	1101	Linalool	6.49	MS, AI, AS
16	1127	*cis*-Limonene oxide	0.14	MS, AI
17	1131	*cis*-p-Mentha-2,8-dien-1-ol	0.20	MS, AI
18	1140	Camphor	0.43	MS, AI
19	1165	ρ-Mentha-1,5-dien-8-ol	0.11	MS, AI
20	1181	meta-Cymen-8-ol	0.11	MS, AI
21	1189	α-Terpineol	0.27	MS, AI
22	1197	*trans*-4-Caranone	0.40	MS, AI
23	1213	*trans*-Carveol	0.14	MS, AI
24	1243	Car-3-en-2-one	0.19	MS, AI
25	1282	Isobornyl acetate	0.16	MS, AI
26	1313	*cis*-2,3-Pinanediol	0.48	MS, AI
27	1323	neoiso-Verbanol acetate	0.17	MS, AI
28	1379	*β*-Patchoulene	0.12	MS, AI
29	1386	(2*E*)-Octenol butanoate	0.15	MS, AI
30	1396	Longifolene	0.13	MS, AI
31	1408	(*Z*)-Caryophyllene	0.17	MS, AI
32	1717	(3*E*)-Butylidene phthalide	1.06	MS, AI
33	1944	*m*-Camphorene	0.29	MS, Nist
34	1980	*p*-Camphorene	0.13	MS, Nist
35	2126	(Z, Z)-9,12-Octadecadienoic acid	0.11	MS, Nist
		Total identified (%)	96.94	

[a] AI = Arithmetic Index on DB5 column. [b] Relative area (peak area relative to the total peak area). [c] MS mass spectrum matching with NIST library; AS, authentic sample.

3.2. Fumigant Toxicity of the Essential Oil

Fumigation assays were conducted to evaluate the fumigant toxicity of essential oils (EOs) against adult *S. zeamais*. The results demonstrated a significant positive correlation between fumigation activity and dosage at 24, 48, and 72 h post-treatment with essential oil (Table 3). At the same dose, corrected mortality also gradually increased with the extension of time, from 24 h to 48 h to 72 h. The largest dose (170 mg/L) of essential oil caused the corrected mortality of 94.48% after 72 h of essential oil treatment (Table 4). The lethal concentration (LC_{50}) values were 164.38, 132.62, and 90.35 mg/L air at 24, 48, and 72 h, respectively.

Table 3. Corrected mortality rate of essential oil of *A. archangelica* against *S. zeamais* adults.

Period (h)	Corrected Mortality Rate (%)				
	34	68	102	136	170
24	6.03 ± 1.06 a	12.19 ± 1.20 b	24.83 ± 1.71 c	38.93 ± 2.46 d	27.72 ± 2.27 e
48	7.53 ± 1.08 a	17.12 ± 1.98 b	32.19 ± 1.67 c	47.26 ± 3.17 b	69.86 ± 3.65 e
72	9.42 ± 1.14 a	23.91 ± 1.62 b	39.96 ± 2.94 c	63.77 ± 4.13 d	93.48 ± 1.35 e

Means in the same row followed by different lowercase letters differed significantly ($p < 0.05$).

Table 4. Fumigant toxicity of essential oil of *A. archangelica* against *S. zeamais* adults.

Period (h)	Toxicity Regression Curves	LC_{50} (mg/L air)	95% FL (mg/L)	Relative Coefficient (R^2)	Chi Square (χ^2)
24	$y = 2.57x - 0.72$	164.38	128.02–296.45	0.99	7.90
48	$y = 2.85x - 1.07$	132.617	104.92–200.42	0.98	9.00
72	$y = 3.11x - 1.04$	90.35	30.60–226.80	0.99	8.07

3.3. Enzyme Activity

The activity of three detoxification enzymes, AchE, GST and CarE of *S. zeamais* was determined. *A. archangelica* essential oil fumigation significantly reduced the activity of AchE, and enzyme activity was negatively correlated with the treatment dosage (Figure 1A). At 24 h after fumigation, the activity of AchE was the lowest, after which it showed an increasing trend but remained lower than that of the control group (Figure 1D). A. archangelica essential oil inhibited the AchE activity in adult *S. zeamais*. The effects of A. archangelica essential oil on GST and AChE in adult *S. zeamais* were generally consistent, both showing an inhibitory effect on enzyme activity. Enzyme activity was negatively correlated with the treatment dosage, and the lowest GST activity was observed at 24 h after fumigation (Figure 1B,E). *A. archangelica* essential oil fumigation significantly reduced CarE activity, with treatment dosage and time having only a minor influence (Figure 1C,F).

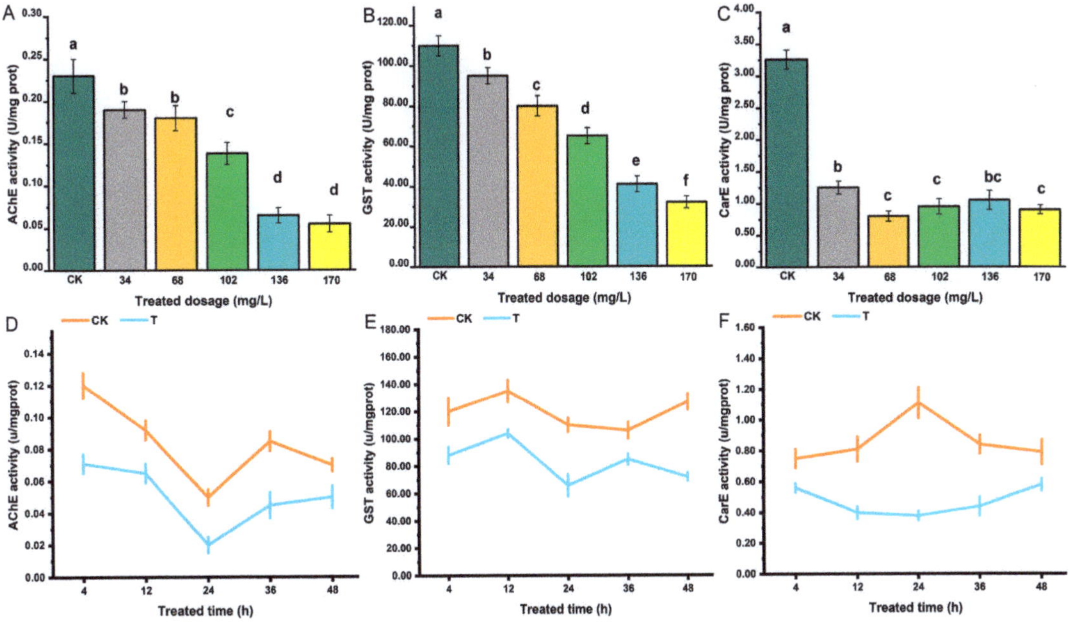

Figure 1. Effects of *A. archangelica* essential oil on the Activities of AChE (**A**,**D**), GST (**B**,**E**) and CarE (**C**,**F**) in *S. zeamais*. (**A–C**) show the effects of essential oil on enzyme activities in 24 h under different dosages; (**D–E**) demonstrate the effects of essential oils on enzyme activities at different time points under the LD50 dosage. Different lowercase letters indicate statistically significant differences among treatments ($p < 0.05$).

3.4. Transcriptome Analysis

RNA-Seq was carried out to explore the differential genes of *S. zeamais* in response to A. archangelica essential oil. As a result, 6.17, 5.99, 6.65 G clean bases for three fumigated groups, and 5.77, 6.75, 5.77 G clean bases for three control groups were produced in this library (Table 5). The percentage of clean reads exceeded 95% of the total reads in each sample. The Q20 and Q30 values of the samples were all above 98% and 95%, respectively, indicating high-quality sequencing data with low error rates and good overall quality. After quality assessment and data filtering, 37,924 unigenes were obtained using Trinity software (Version 2.15.1). Based on the Nr annotation results, we analyzed the species distribution of the annotated sequences. The results revealed that 76.4% of *Sitophilus oryzae* genes matched those in *S. zeamais*, indicating a high degree of genomic similarity between the two species. This was followed by matches to *Rhynchophorus ferrugineus* (2.7%), *Anthonomus grandis grandis* (1.65%), and *Acanthoscelides obtectus* (2%) (Figure 2A). Transcriptomic analysis of *S. zeamais* treated with *A. archangelica* essential oil identified 3718 significantly differentially expressed genes. Expression pattern analysis revealed that 2265 genes were up-regulated, while the remaining 1453 genes were down-regulated (Figure 2B).

Table 5. Summary of statistical data for the transcriptome of *S. zeamais*.

	T1	T2	T3	CK1	CK2	CK3
Raw reads	21,419,493	20,671,023	22,985,184	19,659,294	23,466,508	19,623,581
Raw bases (G)	6.43	6.2	6.9	5.9	7.04	5.89
Clean reads	20,560,357	19,958,351	22,158,451	19,235,246	22,500,960	19,227,394
Clean bases (G)	6.17	5.99	6.65	5.77	6.75	5.77
Clean reads ratio (%)	95.99	96.55	96.40	97.84	95.89	97.98
Q20	98.43	98.55	98.59	98.67	98.61	98.69
GC percentage (%)	36.65	37.54	37.55	43.25	37.33	43.17
Number of unigenes			37,924			

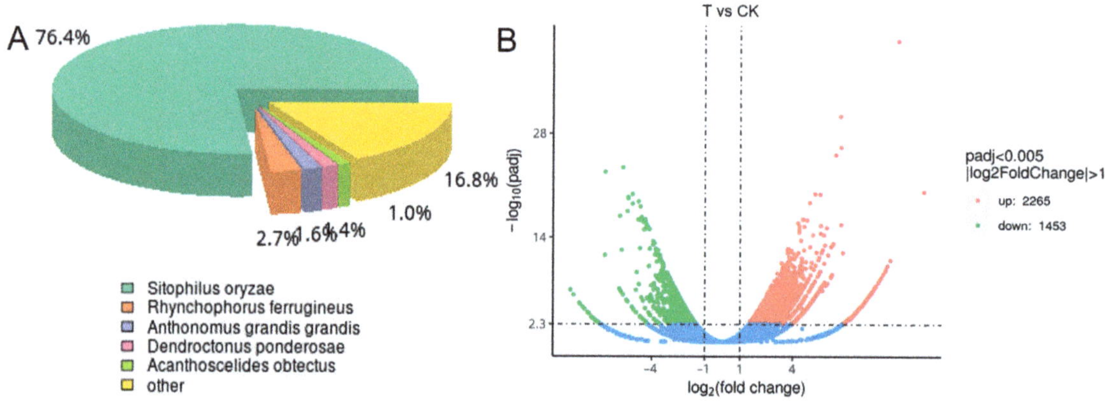

Figure 2. Annotation of species distribution map (**A**) and differential gene volcano map (**B**).

3.5. GO and KEGG Pathway Enrichment Analysis of DEGs

Functional enrichment results indicated that DEGs in GO terms were mainly related to carbohydrate metabolic process, hydrolase activity, oxidoreductase activity, and catalytic activity. They mainly belong to biological processes and molecular functions, respectively (Figure 3A). The DEGs were significantly involved in KEGG pathways including those related to drug metabolism-cytochrome P450, insect hormone biosynthesis, retinol metabolism, terpenoid backbone biosynthesis, glutathione metabolism, fluid shear stress and atherosclerosis, and platinum drug resistance (Figure 3B).

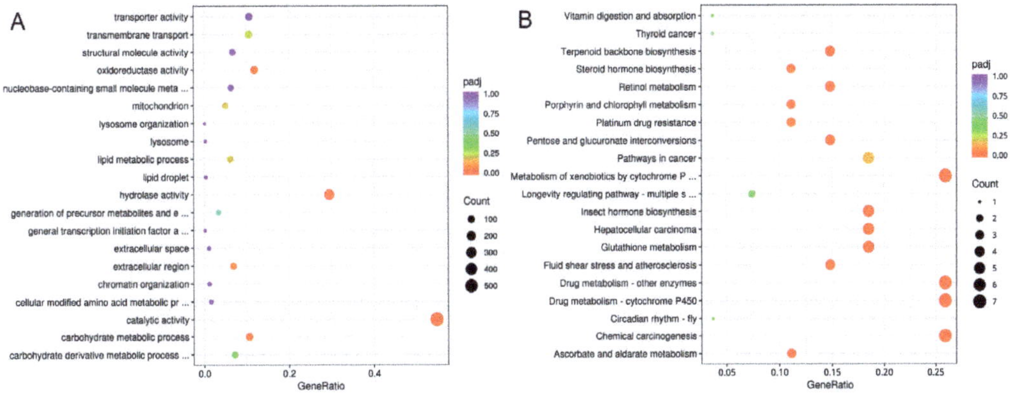

Figure 3. Scatter plot of GO (**A**) and KEGG (**B**) enrichment of DEGs. The horizontal axis represents the ratio of the number of differentially expressed genes annotated to the GO Term/KEGG pathway to the total number of differentially expressed genes, while the vertical axis represents the GO Term and KEGG pathway.

3.6. qRT-PCR Analyses

The expression levels of SzeaCYP6A14, SzeaCYP6M5, SzeaUGT2C1, SzeaUGT2C10, SzeaGST3, and SzeaJHEH1 were significantly higher in the treatment group compared to the control group ($p < 0.05$), whereas SzeaCYP12V2 showed the opposite trend with significantly lower expression in the treatment group ($p < 0.05$) (Figure 4). As expected, the qRT-PCR results for the 14 genes showed similar differential expression patterns between the treatment and control groups as the RNA-seq data, confirming the reliability of our data and analysis.

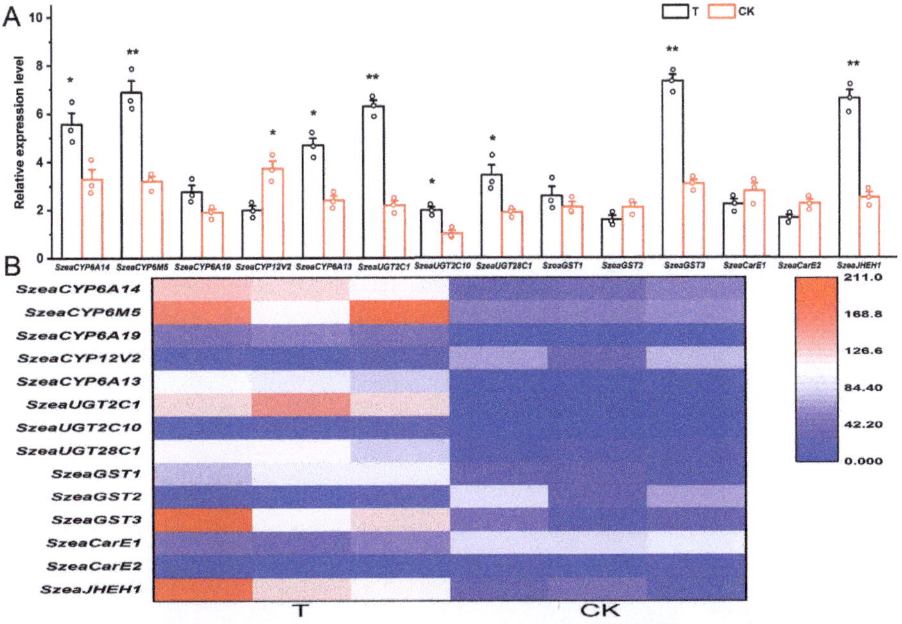

Figure 4. Gene qRT-PCR expression level (**A**) and expression heatmap (**B**). Error bars indicate ± SE. Statistical differences between the treatment and control at the same instar age: *, $p < 0.05$; **, $p < 0.01$.

4. Discussion

Several researchers have examined the chemical constituents of the *A. archangelica* roots essential oil. Fraternale et al. reported the major components of the essential oil were α-pinene (21.3%), δ-3-carene (16.5%), limonene (16.4%) [16]. Wedge et al. also reported α-pinene (17.5%), δ-3-carene (16.3%), limonene (8.5%) [21]. Factors such as variety, geographic and seasonal variations, as well as extraction techniques, may account for the quantitative discrepancies between the values presented in this study and those reported in earlier research.

Angelica essential oils exhibit insecticidal properties against insect pests. The present study is the first to reveal that *A. archangelica* essential oil possesses significant fumigant toxicity against *S. zeamais* adults. Accumulating studies have also reported that the essential oils from other (*Angelica*) species, including *A. pubescens* [22], *A. sinensis* [23], *A. dahurica* [24,25], and *A. archangelica* [26], showed deterrent, repellent, and acute toxicity. Main compounds such as δ-3-Carene, α-pinene, and limonene may be possible mortality of factors for *A. archangelica* essential oil, which are capable of interfering with AChE, CAT and GST activity [27–29]. These enzymes are directly related to the insecticidal mechanism.

Li et al. used *Illicium verum* fruit extracts to fumigate *S. zeamais*. Under LD_{50} treatment at 12, 24, 36, 48, 60, and 72 h, the extract exhibited an inhibitory effect on GST, with the inhibition strengthening over time [30]. Shang et al. demonstrated that after fumigating *S. zeamais* with *Artemisia annua* essential oil, the α-naphthyl acetate esterase (α-NACarE) activity at 2, 4, 8, 12, and 24 h showed a decreasing trend from 2 to 12 h. Although the activity decreased at 2 h and increased again at 4 h, it remained significantly lower than that of the control group [31]. Chaubey et al. reported the effects of α-pinene and β-caryophyllene on acetylcholinesterase (AchE) activity in *S. zeamais*, demonstrating that the enzyme activity in the treated groups decreased significantly. This reveals that α-pinene and β-caryophyllene significantly inhibit AchE activity in *S. zeamais* [32]. These findings are largely consistent with the experimental results obtained in this study. The inhibition of enzyme activity by essential oils may occur because certain small molecules in the oil are structurally similar to the enzyme's natural substrate. These molecules bind to the enzyme's active site but cannot be catalyzed by the enzyme. By occupying the active site, they prevent the genuine substrate from binding to the enzyme, thereby inhibiting its function. The insecticidal mechanism of essential oils may involve a combination of factors, including neurotoxic effects, interference with nerve signal transduction, modulation of GABA receptors, and cellular and physiological toxicity.

Huang et al. evaluated the effects of terpinen-4-ol fumigation on gene expression in *S. zeamais* through RNA-seq analysis, revealing 592 DEGs, among which 308 and 284 genes were upregulated and downregulated, respectively. The DEGs were subjected to GO and KEGG enrichment analyses [9]. KEGG enrichment primarily mapped to metabolic pathways, related to the respiration and metabolism of exogenous organisms [33]. Li et al. extracted total RNA from *S. zeamais* and conducted high-throughput sequencing, reporting GO functional classification and KEGG metabolic pathways [34]. These findings are largely consistent with the experimental results obtained in this study.

5. Conclusions

This study comprehensively evaluated the control efficacy of *A. archangelica* essential oil against *S. zeamais* through GC-MS analysis, fumigation toxicity assays, enzyme activity measurements, RNA-seq sequencing, and qRT-PCR. The research revealed 35 chemical components in the essential oil, its fumigation toxicity to S. zeamais, and its effects on the activities of three detoxification enzymes in the insect. A total of 3718 DEGs were obtained, and their expression patterns were validated. These findings will provide data support for developing novel, eco-friendly, plant-based pesticides.

Author Contributions: Conceptualization, G.W. and H.W.; methodology, G.W., D.L., Z.Z. and X.G.; software, G.W., L.W., S.S. and X.J.; validation, Z.Z., D.L. and X.G.; formal analysis, L.W., S.S. and X.J.; investigation, K.K. and S.D.; resources, B.Z., S.D. and H.W.; data curation, Z.Z., G.W., D.L. and X.G.; writing—original draft preparation, G.W. and D.L.; writing—review and editing, X.G. and H.W.; visualization, D.L.; supervision, B.Z., K.K. and S.D.; project administration, G.W. and X.G.; funding acquisition, H.W. All authors have read and agreed to the published version of the manuscript.

Funding: This research was funded by the Key Scientific and Technological Research Project of Henan Province (No. 252102111108 and 242102110194), the Key Research and Development and Promotion Projects of Henan Province (No. 241111111700).

Institutional Review Board Statement: Not applicable.

Informed Consent Statement: Not applicable.

Data Availability Statement: All the associated data are available in the manuscript.

Acknowledgments: We thank the anonymous reviewers for valuable comments on the manuscript. We are grateful for the assistance of all staff members and students in the Laboratory of Entomology and Pesticide, Henan Institute of Science and Technology, Xinxiang, Henan, China.

Conflicts of Interest: The authors declare no conflicts of interest.

References

1. Kifle, F.; Girma, M.; Gebresilassie, A.; Woldehawariat, Y.; Ele, E. Chemical composition and insecticidal potential of botanical fractionation extracts for the management of *Sitophilus zeamais* Motschulsky, 1855 (Coleoptera: Curculionidae) in stored maize. *Helicon* **2025**, *11*, e42131. [CrossRef] [PubMed]
2. Wanna, R.; Bunphan, D.; Kunlanit, B.; Khaengkhan, P.; Khaengkhan, P.; Bozdoğan, H. Chemical composition of essential oil from *Apium graveolens* L. and its biological activities against *Sitophilus zeamais* Motschulsky (Coleoptera: Dryophthoridae). *Plants* **2025**, *14*, 347. [CrossRef]
3. Usseglio, V.; Dambolena, J.S.; Zunino, M.P. Insect-corn kernel interaction: Chemical signaling of the grain and host recognition by *Sitophilus zeamais*. *J. Stored Prod. Res.* **2018**, *79*, 66–72. [CrossRef]
4. Boyer, S.; Zhang, H.; Lempérière, G. A review of control methods and resistance mechanisms in stored-product insects. *Bull. Entomol. Res.* **2012**, *102*, 213–229. [CrossRef] [PubMed]
5. Liao, M.; Xiao, J.J.; Zhou, L.J.; Liu, Y.; Wu, X.W.; Hua, R.M.; Wang, G.R.; Cao, H.Q. Insecticidal activity of melaleuca alternifolia essential oil and RNA-Seq analysis of Sitophilus *zeamais* transcriptome in response to oil fumigation. *PLoS ONE* **2016**, *11*, e0167748. [CrossRef]
6. Nagawa, C.B.; Kitiibwa, I.S.; Kizito, S.S.; Syofuna, A.; Kyarimpa, C.M.; Omara, T. Chemical composition and insecticidal potential of *Eucalyptus* essential oils against *Sitophilus zeamais* (Motschulsky, 1855) and *Acanthoscelides obtectus* (Say, 1831). *S. Afr. J. Bot.* **2025**, *179*, 48–55. [CrossRef]
7. Wanna, R.; Kaewduangta, W. Fumigant activity of *Tridax procumbens* (Asterales: Asteraceae) essential oil against *Sitophilus zeamais* (Coleoptera: Curculionidae) and its effects on Thai rice seed germination. *J. Entomol. Sci.* **2022**, *57*, 561–572. [CrossRef]
8. Liao, M.; Li, S.N.; Wu, H.L.; Gao, Q.; Shi, S.; Huang, Y.; Cao, H.Q. Transcriptomic analysis of *Sitophilus zeamais* in response to limonene fumigation. *Pest Manag. Sci.* **2022**, *78*, 4774–4782. [CrossRef]
9. Huang, Y.; Liao, M.; Yang, Q.Q.; Xiao, J.J.; Hu, Z.Y.; Zhou, L.J.; Cao, H.Q. Transcriptome profiling reveals differential gene expression of detoxification enzymes in *Sitophilus zeamais* responding to terpinen-4-ol fumigation. *Pestic. Biochem. Physiol.* **2018**, *149*, 44–53. [CrossRef]
10. Bounouira, Y.; Benyelles, N.G.; Senouci, H.; Benazzouz, F.Z.; Chaieb, I. The insecticidal activity of a formulation of ammoides verticillata essential oil and diatomaceous earth on *Sitophilus zeamais*. *Int. J. Trop. Insect. Sci.* **2022**, *42*, 2979–2985. [CrossRef]
11. Aryal, S.; Poudel, A.; Kafle, K.; Aryal, L.N. Insecticidal toxicity of essential oil of *Nepalese Acorus calamus* (Acorales: Acoraceae) against *Sitophilus zeamais* (Coleoptera: Curculionidae). *Heliyon* **2023**, *9*, e22130. [CrossRef] [PubMed]
12. Fouad, H.A.; Câmara, C.A.G.; Moraes, M.M.; Tavares, W.S.; Legaspi, J.C.; Zanuncio, J.C. Insecticidal and repellent activities of four essential oils against *Sitophilus zeamais* (Coleoptera: Curculionidae). *Dose Response* **2023**, *21*, 15593258231210263. [CrossRef]

13. Rodríguez, A.; Beato, M.; Usseglio, V.L.; Camina, J.L.; Zygadlo, J.A.; Dambolena, J.S.; Zunino, M.P. Phenolic compounds as controllers of *Sitophilus zeamais*: A look at the structure-activity relationship. *J. Stored Prod. Res.* **2022**, *99*, 102038. [CrossRef]
14. Pathak, S.; Wanjari, M.M.; Jain, S.K.; Tripathi, M. Evaluation of antiseizure activity of essential oil from roots of *Angelica archangelica* Linn. in mice. *Indian J. Pharm. Sci.* **2010**, *73*, 371–375.
15. Aćimović, M.G.; Pavlović, S.D.; Varga, A.O.; Filipović, V.M.; Cvetković, M.T.; Stanković, J.M.; Čabarkapa, I.S. Chemical composition and antibacterial activity of *Angelica archangelica* root essential oil. *Nat. Prod. Commun.* **2017**, *12*, 205–206. [CrossRef]
16. Fraternale, D.; Flamini, G.; Ricci, D. Essential oil composition and antimicrobial activity of *Angelica archangelica* L. (Apiaceae) roots. *J. Med. Food* **2014**, *17*, 1043–1047. [CrossRef]
17. Grabherr, M.G.; Haas, B.J.; Yassour, M.; Levin, J.Z.; Thompson, D.A.; Amit, I.; Adiconis, X.; Fan, L.; Raychowdhury, R.; Zeng, Q.D.; et al. Full-length transcriptome assembly from RNA-Seq data without a reference genome. *Nat. Biotechnol.* **2011**, *29*, 644–652. [CrossRef]
18. Li, D.M.; Zand, M.S.; Dye, T.D.; Goniewicz, M.L.; Rahman, I.; Xie, Z.D. An evaluation of RNA-seq differential analysis methods. *PLoS ONE* **2022**, *17*, e0264246. [CrossRef] [PubMed]
19. Lü, D.B.; Yan, Z.Z.; Hu, D.; Zhao, A.P.; Wei, S.J.; Wang, P.; Yuan, X.Q.; Li, Y.P. RNA Sequencing reveals the potential adaptation mechanism to different hosts of *Grapholita molesta*. *Insects* **2022**, *13*, 893. [CrossRef] [PubMed]
20. Prentice, K.; Pertry, I.; Christiaens, O.; Bauters, L.; Bailey, A.; Niblett, C.; Ghislain, M.; Gheysen, G.; Smagghe, G. Transcriptome analysis and systemic RNAi response in the African sweetpotato weevil (*Cylas puncticollis*, Coleoptera, Brentidae). *PLoS ONE* **2015**, *10*, e0115336. [CrossRef]
21. Wedge, D.E.; Klun, J.A.; Tabanca, N.; Demirci, B.; Ozek, T.; Baser, K.H.C.; Liu, Z.; Zhang, S.; Cantrell, C.L.; Zhang, J. Bioactivity-guided fractionation and GC/MS fingerprinting of Angelica sinensis and *Angelica archangelica* root components for antifungal and mosquito deterrent activity. *J. Agric. Food Chem.* **2009**, *57*, 464–470. [CrossRef]
22. Zheng, R.R.; Zhao, J.Y.; Ma, L.; Qie, X.T.; Yan, X.Z.; Hao, C. Behavioral, electrophysiological, and toxicological responses of *Plutella xylostella* to extracts from *Angelica pubescens*. *Insects* **2023**, *14*, 613. [CrossRef]
23. Champakaew, D.; Junkum, A.; Chaithong, U.; Jitpakdi, A.; Riyong, D.; Sanghong, R.; Pitasawat, B. *Angelica sinensis* (Umbelliferae) with proven repellent properties against *Aedes aegypti*, the primary dengue fever vector in Thailand. *Parasitol. Res.* **2015**, *114*, 2187–2198. [CrossRef] [PubMed]
24. Tabanca, N.; Gao, Z.; Demirci, B.; Techen, N.; Wedge, D.E.; Ali, A.; Baser, K.H.C. Molecular and phytochemical investigation of *Angelica dahurica* and *Angelica pubescentis* essential oils and their biological activity against *Aedes aegypti*, *Stephanitis pyrioides*, and *Colletotrichum* species. *J. Agric. Food Chem.* **2014**, *62*, 8848–8857. [CrossRef]
25. Chung, I.M.; Kim, E.H.; Lee, J.H.; Lee, Y.C.; Moon, H.I. Immunotoxicity activity from various essential oils of *Angelica genus* from South Korea against *Aedes aegypti* L. *Immunopharm. Immunot.* **2012**, *34*, 42–45.
26. Pavela, R.; Vrchotová, N. Insecticidal effect of furanocoumarins from fruits of *Angelica archangelica* L. against larvae *Spodoptera littoralis* Boisd. *Ind. Crop Prod.* **2013**, *43*, 33–39. [CrossRef]
27. Galeano, L.J.N.; Prieto-Rodríguez, J.A.; Patiño-Ladino, O.J. Synergistic insecticidal activity of plant volatile compounds: Impact on neurotransmission and detoxification enzymes in *Sitophilus zeamais*. *Insects* **2025**, *16*, 609. [CrossRef] [PubMed]
28. de Carvalho Brito, R.; de Moura Pádua, L.E.; da Silva, L.R.; Briozo, M.E.O.; Silva, P.R.R.; de Carvalho, L.F.; de França, S.M. Use of essential oils and α-pinene as insecticides against *Sitophilus zeamais* and their effects on maize seed germination. *Agronomy* **2024**, *14*, 2282. [CrossRef]
29. Chaubey, M.K. Insecticidal property of terpenes against maize weevil, *Sitophilus zeamais* (Motschulsky). *J. Biopestic.* **2022**, *15*, 92. [CrossRef]
30. Li, S.G.; Li, M.Y.; Huang, Y.Z.; Hua, R.M.; Lin, H.F.; He, Y.J.; Wei, L.L.; Liu, Z.Q. Fumigant activity of Illicium verum fruit extracts and their effects on the acetylcholinesterase and glutathione S-transferase activities in adult *Sitophilus zeamais*. *J. Pest Sci.* **2013**, *86*, 677–683. [CrossRef]
31. Shang, L.N.; Yuan, H.B.; Wei, C.Y.; Ren, B.Z. Effect of essential oil extracted from *Artemisia annua* L. on activities of several enzymes in *Sitophihus zeamais*. *J. Jilin Agr. Univ.* **2010**, *32*, 616–621.
32. Chaubey, M.K. Essential oils as green pesticides of stored grain insects. *Eur. J. Bio. Res.* **2019**, *4*, 202–244.
33. Bens, M.; Sahm, A.; Groth, M.; Jahn, N.; Morhart, M.; Holtze, S.; Hildebrandt, T.B.; Platzer, M.; Szafranski, K. FRAMA: From RNA-seq data to annotated mRNA assemblies. *BMC Genom.* **2016**, *17*, 54. [CrossRef] [PubMed]
34. Li, Y.Y.; Zhao, J.H. Transcriptome analysis of Sitophilus zeamais by RNA-Seq. *J. Environ. Entomol.* **2021**, *43*, 1485–1492.

biology

MDPI

Article

Behavioral Responses of the Bumblebee *Bombus terrestris* to Volatile Compounds from Blueberries

Yun He [1], Jiaru Zhang [1], Ziyang Hu [1], Yingxue Cao [1], Kevin H. Mayo [2], Duo Liu [1,*] and Mingju E [1,*]

[1] School of Life Sciences, Changchun Normal University, 677 Changjibei Road, Changchun 130032, China; qx202310015@stu.ccsfu.edu.cn (Y.H.)

[2] Department of Biochemistry, Molecular Biology & Biophysics, University of Minnesota Health Sciences Center, 6-155 Jackson Hall, Minneapolis, MN 55455, USA

* Correspondence: liuduoccsfu@163.com (D.L.); emj640@nenu.edu.cn (M.E.)

Simple Summary: Blueberries rely on insects to carry pollen between flowers, but natural pollination is often too low to ensure high fruit yields. This study explored which floral scents attract buff-tailed bumblebees, the most effective pollinators of this crop. We collected air samples from around blueberry blossoms and found 32 different fragrance compounds, with two scents, linalool and styrene, being the most common. We then tested how the bees reacted to these scents by measuring their antennal signals and observing their choices in a simple maze. The bees were drawn to low levels of certain sweet, plant-based alcohols, while high amounts of some sharp-smelling chemicals turned them away. These results identify the key scents that guide bees to flowers and open the door to creating natural sprays that can safely draw more pollinators to blueberry fields. Such sprays could help farmers produce larger harvests and higher-quality fruit without relying on harmful chemicals, supporting both agriculture and the environment.

Abstract: Bumblebees are important pollinating insects in crop pollination. Chemical attractants can effectively improve the flower-visiting efficiency of bumblebees, thereby increasing blueberry yields. To identify volatile compounds that attract bumblebees, we collected volatile compounds from blueberry flowers using headspace extraction. Gas chromatography– mass spectrometry (GC–MS) identified 32 volatile compounds, with Linalool and Styrene being the primary substances that accounted for 25.93% and 14.28%, respectively. The olfactory threshold of bumblebee antennae was assessed using electroantennography (EAG), and the behavioral responses from bumblebees were investigated using a Y-tube olfactometer. Results indicate that among the six classes of volatiles—alcohols, aldehydes, esters, ketones, aromatic compounds, and olefins—alcohols constituted the predominant proportion. Among these, six compounds—benzaldehyde, phenylpropylaldehyde, citral, linalool, α-terpineol, and geraniol—induced significant antennal responses in bumblebees. Our assays showed that geraniol, linalool, and α-terpineol at concentrations of 0.1 μg/μL, 1 μg/μL, and 10 μg/μL elicited attraction, whereas higher concentrations of benzaldehyde, benzenepropanal, and citral had repellent effects.

Keywords: *Bombus terrestris*; *Vaccinium* spp.; volatile compounds; EAG

1. Introduction

Bumblebees (*Bombus terrestris*), key pollinators, are large, cold-tolerant and densely furred, with excellent low-light adaptability, rapid flower-visiting speed and vigorous foraging capacity [1,2]. Their ability to perform buzz-pollination—sonication of poricidal anthers—makes them especially effective for crops such as tomato and blueberry, where pollen release otherwise limits fruit production. Bumblebees are now the standard managed pollinator for greenhouse tomatoes, peppers, strawberries, pumpkins and blueberries [3,4], and their deployment has reduced labor costs of hand pollination while increasing yield stability across multiple continents.

In the pollinator environment, most olfactory signals are volatile compounds—plant volatiles, sex pheromones, alarm pheromones, etc. [5–7]—synthesized by all plant organs [8] and strongly influencing insect behavior [9]. Floral volatiles not only attract pollinators but also shape their nectar-collecting choices; for example, specific aromatics increase bees' preference for high-sugar flowers [10]. For bumblebees, aromatic or relatively strong odors are particularly attractive: E-β-bergamotene emitted by *Oxalis corniculata* L. [11] and phenylacetaldehyde from rapeseed (*Brassica rapa* L.) [12] significantly enhance visitation. Yet which volatiles mediate close-range attraction of bumblebees to blueberry flowers remains unclear.

Blueberry (*Vaccinium* spp.) is prized for its delicate flesh, antioxidant content and premium price [13–15], but self-pollination stands at <30% of flowers and open-field fruit pollination depends almost entirely on insects [16,17]. Field studies show that a single visit by a buzzing *B. terrestris* increases fertilization success by >30%, berry weight by 8–12%, seed number by 20–40% and the share of export-grade fruit by 15–20%, translating into increased fruit gains of 20–35% and simultaneous reductions in malformation and firmness loss [18,19]. Consequently, bumblebee visitation is the primary determinant for both volume and commercial quality for blueberry harvests.

This study investigated blueberry floral volatiles using solid-phase microextraction (SPME), identified the components by GC–MS and screened electrophysiologically active compounds using GC–EAG on bumblebee antennae. Behavioral assays were then used to quantify bee preferences for individual volatiles, with the aim of identifying the chemical basis of the blueberry–Bombus interaction and developing natural pollinator attractants to improve pollination rates, yield and berry quality.

2. Materials and Methods

2.1. Insects

Healthy female bumblebees used here were obtained from the Jilin City Bee Research Institute in Jilin Province and kept in insect rearing cages (18 cm × 18 cm × 10 cm). The lab temperature was set at 26 ± 1 °C in a red-light environment with relative humidity maintained at 60 ± 10%. Bumblebees were fed a solution of pollen and sucrose (concentration ratio 1:1.2).

2.2. Collection and Analysis of Headspace Volatiles from *Vaccinium* spp.

In early May 2024, during peak blueberry flowering season, we acquired samples from three blueberry plantations across Changbai Mountain at the outdoor base in Guanghua Town, Tonghua County, Baishan City, Jilin Province. From these samples, we obtained a comprehensive volatile compound profile. Nine plants (three per site) with uniform vegetative growth were selected, and 8–12 freshly opened flowers were picked from each plant. Ambient conditions at sampling averaged 65 ± 5% RH and 18 ± 2 °C, typical for early May in the Changbai mountain range. One week after anthesis, intact flowers were cut with tweezers, immediately placed in 20 mL headspace vials, and subjected to SPME (CAR/DVB/PDMS fiber) extraction at 70 °C for 30 min; the fiber was then desorbed in the GC–MS injector at 250 °C for 5 min. Each set of flowers from a single plant constituted one replicate, giving a total of nine replicates (three plants × three sites).

2.3. Gas Chromatography–Mass Spectrometry (GC–MS)

Extracts were analyzed by GC–MS using an HP-5MS flexible quartz capillary column (30 m × 0.25 mm × 0.25 μm Agilent Technologies, Santa Clara, CA, USA) equipped with an Agilent 7000C. The carrier gas was high-purity helium (purity ≥ 99.99%) at a flow rate of 1 mL/min in splitless mode, with an injection port temperature of 250 °C. Column oven temperature program settings were as follows: maintain at 40 °C for 3 min, then increase the temperature at a rate of 10 °C/min to 230 °C and maintain it for 5 min. We used the full scan acquisition mode, scanning range of *m/z* 30–550, ion source temperature 230 °C, transfer port temperature 280 °C, and quadrupole temperature 180 °C. We identified compounds based on the NIST14 spectral library, then we used standard samples to determine whether the retention times were consistent. We then calculated the relative mass fraction using the Area Normalisation Method [20].

2.4. Electroantennography (EAG)

Paraffin oil was used as the solvent to prepare solutions for each standard compound at concentrations of 0.01, 0.1, 1, 10, and 100 μg/μL. When not in use, the solutions were stored at −20 °C. Liquid paraffin was used as a control. All chemicals were purchased from Changchun Landel Technology Co., Ltd., Changchun, China. In preliminary tests, the lowest (0.01 μg/μL) and highest (100 μg/μL) concentrations were found to mark the response thresholds of bumblebees to the volatiles. Using EAG, volatile substances with a concentration of 1 μg/μL were initially screened, and then dose–response experiments were conducted on volatile substances that produced a strong electrical response in bumblebees during the initial screening. Before testing, adult bumblebees were starved for 3 h.

We applied an appropriate amount of conductive adhesive (SpectraR360, PPG Industries, Pittsburgh, PA, USA) to the EAG probe head and used a scalpel (Feather Safety Razor Co., Osaka, Japan) to cut off the pair of bumblebee antennae along the base, removed the scape, and then removed the front end of the antennal flagellum. We connected the antennae to the electrode coated with conductive adhesive, adjusted the position to ensure that both ends of the antennae were in full contact with the electrode surface, and we made sure that the antennae were not bent when detached from the body. The electroantennography air pump sent two test air streams to blow on the detached antennae simultaneously. One stream was a continuous air stream that had been filtered and humidified before being blown out, and the other stream was an air stream carrying the sample stimulus odor.

During testing, we took 10 μL of paraffin oil with the sample to be tested, mixed them evenly, and added them to the center of a 4 cm long × 0.2 cm wide strip of filter paper that was cut in advance. We then placed it in a Pasteur tube (paraffin oil is used to balance the odor). Before testing, we used a piece of hard cardboard to stimulate the detached antennae to detect their activity level. If there were significant fluctuations, we allowed the test to proceed. We waited until a stable baseline appeared on the recorder and used the foot pedal to trigger the stimulus.

The stimulus duration time was ~0.3 s, with intervals of more than 30 s between stimuli to allow the antennae's sensory function to fully recover and minimize error. The stimulation sequence alternated between paraffin oil and the sample. The sample-testing sequence progressed from low to high concentration. Each sample was tested 4 times, and the test antennae were replaced after each sample was tested once. The test antennae were cut as needed to maintain their activity. Since the EAG response of the insect was relatively stable in the morning, all experiments were conducted between 7:00 and 11:00 a.m. The

EAG reaction test results are expressed as the relative EAG reaction value of worker bees to the sample (m represents the Mean), as shown in Equation (1):

$$\text{EAG relative response value} = (\text{m(Sample stimulation)} - \text{m(Contrast stimulation)})/(\text{m(Contrast stimulation)}) \times 100\% \quad (1)$$

2.5. Y-Tube Olfactometer Choice-Behavioral Assays

From the EAG reaction test results, volatile compounds with a significant stimulating effect on bumblebees were selected for behavioral assays. All 60 worker bees tested were randomly drawn from 15 colonies (ca. 300 bees each). The Y-tube had a main arm length of 24 cm, side arms of 26 cm angled at 60°, with an inner diameter of 4 cm for both the main arm and side arms. The two arms were connected to the odor source bottle and the control bottle, respectively. Before the airflow (200 mL min^{-1}) entered the odor source bottle and the control bottle, it was first passed through an activated carbon filter and a distilled water bottle to purify the air and increase the air humidity. We placed the test bumblebees in the middle of the Y-tube, one bumblebee at a time, and observed the bumblebees' preferences for different odors and concentrations. We replaced the sample and control paraffin oil for every 10 bumblebees. Each concentration of an odor was tested on 60 bumblebees. The evaluation criteria were as follows: each worker bee was observed for a minimum of 2 min, and the longest time spent by a bumblebee entering the insect odor source bottle or remaining within the 1/3 area of the odor source bottle was considered a selection of that substance. When changing different odor sources, we wiped the inner and outer walls of the Y-tube with ultrapure water, allowed it to air dry, and only continued the test after no odor residue remained. The test duration was the same as in the EAG test, both conducted between 7:00 and 11:00 a.m.

2.6. Data Analysis

All statistical analyses and graphical representations were performed using GraphPad Prism 9 (Company: GraphPad Software Inc., San Diego, CA, USA). For the analysis of EAG response results, we used the one-way analysis of variance (ANOVA) and multiple comparisons to analyze significant differences. Behavioral responses were analyzed using the chi-square test. Differences were considered significant with $p < 0.05$.

3. Results

3.1. Analysis of the Types and Contents of Volatiles

The analysis of the types and contents of volatiles from GC–MS was employed to identify compounds compared to those in the NIST14 spectral library. Quantitative analysis was performed using the normalized total ion current peak area. As shown in Table 1, volatile compounds released by blueberry flowers belong to six categories and comprise 32 components, including 10 alcohols, 5 aldehydes, 4 esters, 1 ketone, 3 aromatic compounds, and 9 olefins. Among these, linalool and styrene were the primary components, with relative amounts of 25.93% and 14.28%, respectively.

3.2. EAG Responses

Among the 32 compounds detected in the airborne odor of blueberry flowers, only six elicited strong antennal electrical responses in bumblebees in the EAG experiment. Six compounds—benzaldehyde, phenylpropylaldehyde, citral, linalool, α-terpineol, and geraniol (Figure 1)—were identified as the key active constituents responsible for attracting pollinators. Among them, benzaldehyde caused a higher bumblebee antennal electrical response than geraniol.

Table 1. Volatile compounds of *Vaccinium* spp.

CM	Compounds	CAS Number	Retain Time (s)	Relative Content (%)
Alcohols				
	Benzyl alcohol	100-51-6	9.028	0.92
	Linalool	78-70-6	10.419	25.93
	4-Carvomenthenol	562-74-3	11.727	2.36
	2-(4-Methylphenyl)propan-2-ol	1197-01-9	11.848	1.06
	α-Terpineol	98-55-5	11.948	3.92
	(R)-(+)-β-Citronello	1117-61-9	12.511	6.07
	Geraniol	106-24-1	12.923	5.78
	2-Undecanol	1653-30-1	13.569	3.32
	(E)-Cinnamyl Alcohol	4407-36-7	13.702	0.42
	(E)-3,7,11-Trimethyldodeca-1,6,10-trien-3-ol	40716-66-3	16.985	0.42
Aldehydes				
	Benzaldehyde	100-52-7	7.153	1.79
	Phenylacetaldehyde	122-78-1	9.194	1.64
	Phenylpropylaldehyde	104-53-0	11.461	0.64
	Decanal	112-31-2	12.123	0.43
	Citral	5392-40-5	12.702	0.58
Esters				
	cis-3-Hexenyl acetate	3681-71-8	8.424	1.20
	Formic acid, octyl ester	112-32-3	9.803	1.26
	Benzoic acid, ethyl ester	93-89-0	11.598	6.21
	cis-3-Hexenyl 3-methylbutanoate	35154-45-1	12.59	1.80
Ketones				
	2-Undecanone	112-12-9	13.452	2.06
Aromatic compounds				
	m-Cymene	535-77-3	8.769	2.92
	Eugenol	97-53-0	14.406	2.89
	methoxy-4-(1-propenyl)-Phenol	5912-86-7	15.631	0.46
Olefins				
	Styrene	100-42-5	4.812	14.28
	β-Myrcene	123-35-3	8.024	2.09
	Dipentene	138-86-3	8.857	3.87
	α-Cubebene	17699-14-8	14.319	0.44
	α-copaene	3856-25-5	14.706	0.99
	(-)-β-Bourbonene	5208-59-3	14.844	0.75
	β-Caryophyllene	87-44-5	15.319	1.61
	α-Caryophyllene	6753-98-6	15.761	0.81
	γ-Muurolene	30021-74-0	16.015	0.29

For most test stimuli, the antennal response increased with increasing compound concentration. However, the thresholds for antennal electrical responses to different volatiles varied from bumblebee to bumblebee. Antennal responses to benzaldehyde and α-terpineol reached the maximum at 100 μg/μL, whereas sensitivity at 10 μg/μL was greater for phenylpropylaldehyde, citral, linalool, and geraniol (Figure 2).

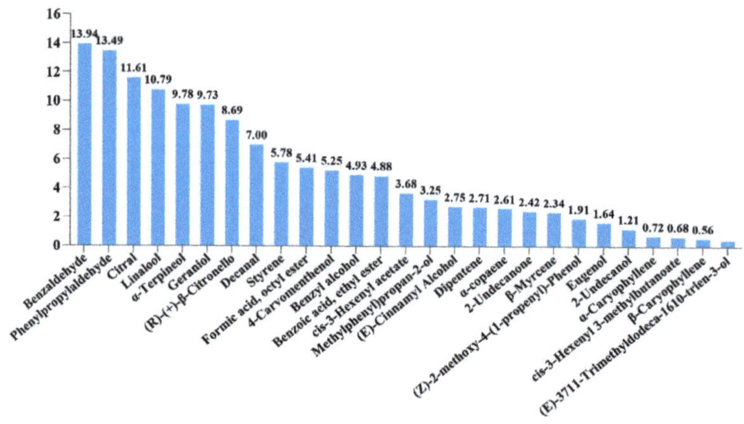

Figure 1. EAG relative response values of each volatile of *Vaccinium* spp. at a concentration of 1 μg/μL.

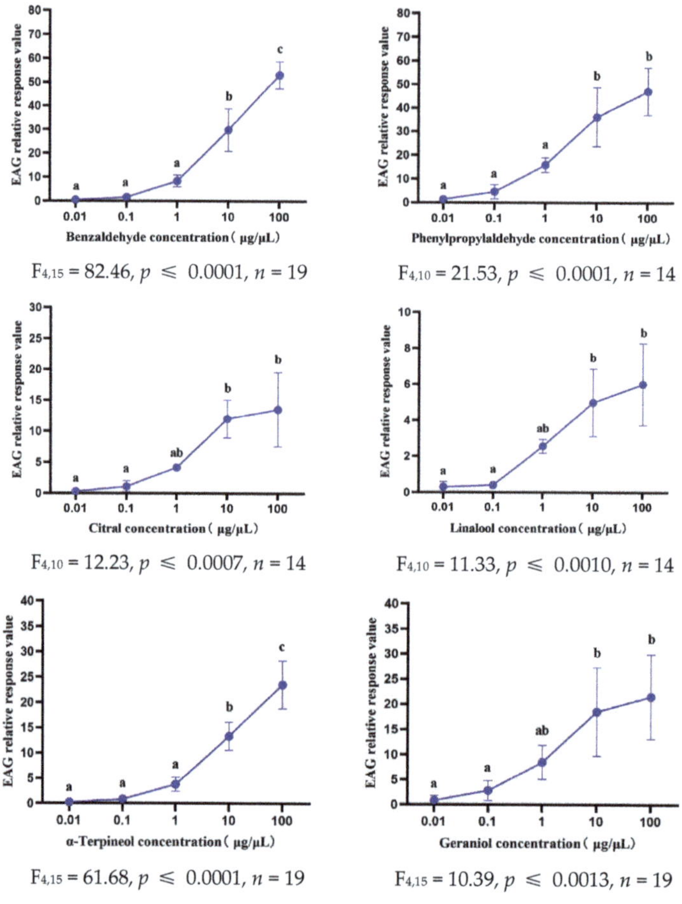

Figure 2. Dose–response of EAG active volatiles. Perform a one-way ANOVA analysis on the EAG response results. Significant differences are indicated by lowercase letters: the same letter indicates no difference, and different letters indicate significant differences $p < 0.05$.

3.3. Choice Behavioral Responses

Six volatile compounds that elicited strong antennal electrical responses in bumblebees were used in olfactory and choice behavioral experiments. Bumblebees exhibited clear concentration-dependent responses to the six EAG-activating compounds tested. Overall, geraniol had a significant attractiveness to bumblebees at 0.1, 1, and 10 μg/μL (Figure 3b–d) and linalool only exhibited an attractive effect at 1 μg/μL (Figure 3c). However, α-terpineol exhibited attractiveness at 0.1 and 1 μg/μL, but repelled bumblebees at 100 μg/μL (Figure 3b,c,e). Phenylacetaldehyde repelled bumblebees at 1, 10, and 100 μg/μL (Figure 3c–e), and citral was repellent to bumblebees at 10 and 100 μg/μL (Figure 3d,e). However, benzaldehyde only caused bumblebees to exhibit clear repellency at 10 μg/μL (Figure 3d). In addition, the concentration of volatile repulsive compounds was generally higher than that of volatile attractive compounds (Figure 3).

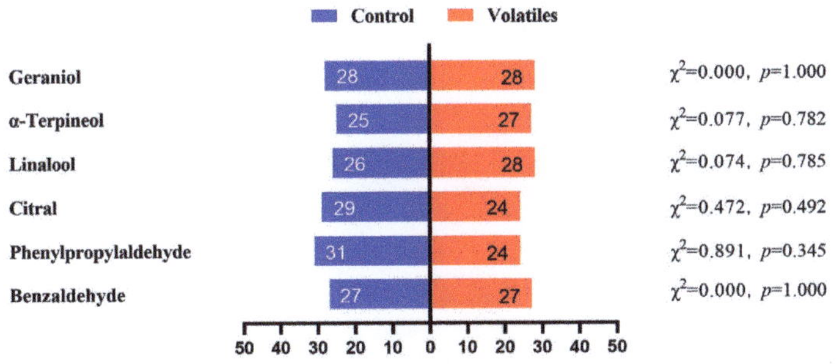

(a)

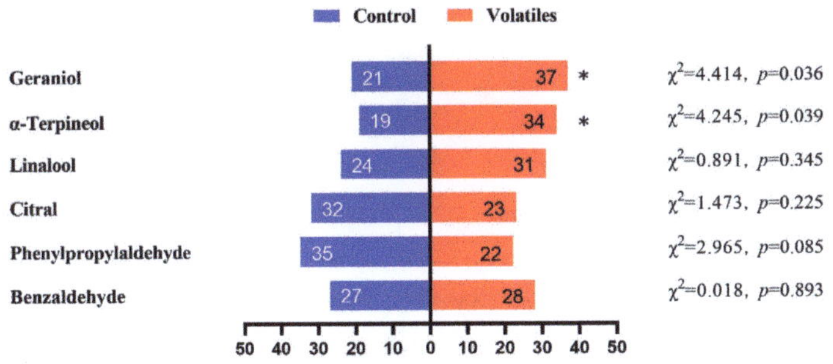

(b)

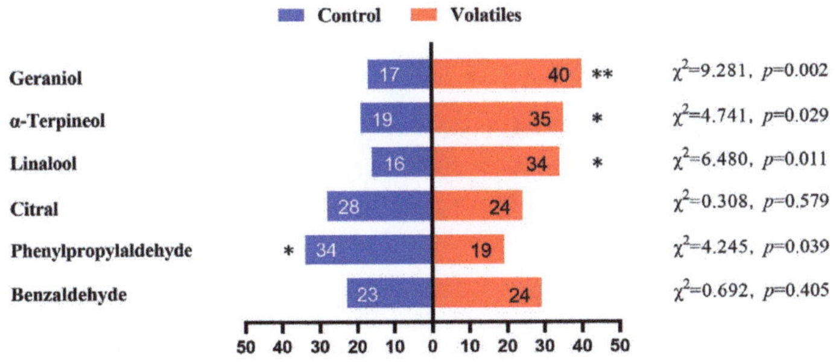

(c)

Figure 3. *Cont.*

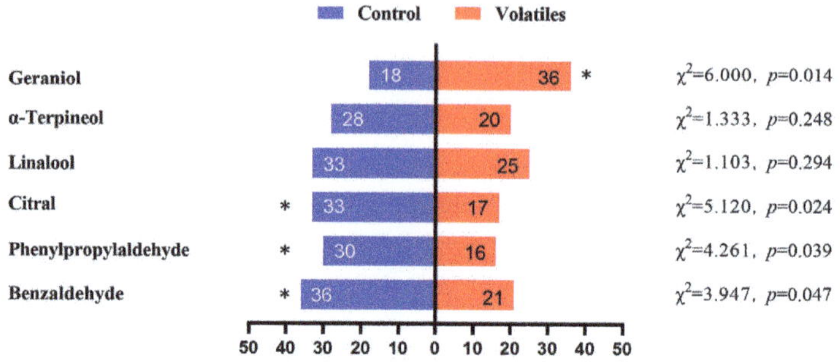

(d)

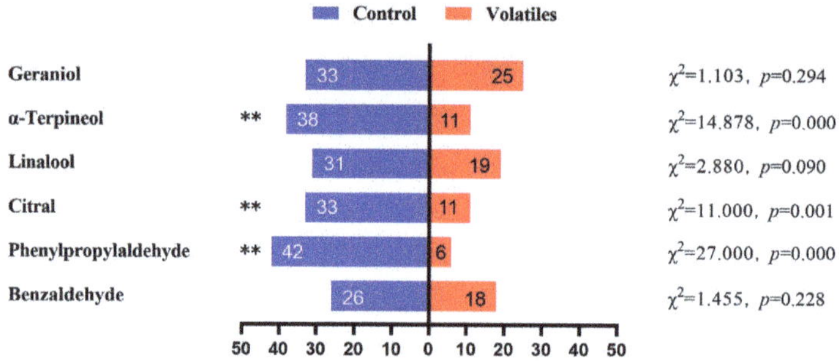

(e)

Figure 3. Behavioral responses of bumblebees to six volatiles. The corresponding volatile concentrations for (**a–e**) are 0.01 μg/μL, 0.1 μg/μL, 1 μg/μL, 10 μg/μL, and 100 μg/μL. The right side shows six volatile substances, while the left side shows the control (paraffin oil). Positive values on the right represent the attraction effect, and negative values on the left represent avoidance. The numbers in each bar chart represent the number of bumblebees that made the selection. In the chi-square test, * indicates a significant difference between volatile and control groups with $p < 0.05$, ** indicates a significant difference between the two with $p < 0.01$.

4. Discussion

In nature, floral scent—composed of species-specific blends of volatile organic compounds (VOCs) released as secondary metabolites—acts together with visual cues to attract or orient pollinators [21–23]. These odors span fragrant and malodorous types, both of which promote pollination: sweet aromas enable bees and butterflies to locate nectar rewards, whereas fetid emissions attract flies or beetles [24–27]. Studies have demonstrated that bumblebees prefer brightly colored, aromatic flowers such as melon, mimosa, clematis, and tomato [28–30], underscoring the pivotal role of floral volatiles in mediating plant–pollinator interactions.

In our study, we employed GC–MS, EAG, and behavioral experiments to determine the electrophysiological and behavioral activity of blueberry floral volatiles on bumblebees. The active compounds identified are generally alcohols, aldehydes, esters, ketones, aromatic compounds, and olefins. Here, we detected six EAG-active compounds, including benzaldehyde, phenylpropyl aldehyde, citral, linalool, α-terpineol, and geraniol. Five volatile substances with significant effects were identified, i.e., linalool, cinnamyl alcohol, 1-pentene-3-ol, hexanal, and (E)-2-hexenal [31–33]. The reason for minimal chemical similarity may be due to differences in blueberry varieties and growing regions. A study by Rodriguez-Saona et al. showed that changes in blueberry flower volatiles depend on various factors, including pollination status, plant cultivation, and the time of day [34].

Another possibility is that discrepancies may be influenced by differences in cultivation, temperature, humidity, soil properties, analytical conditions employed, and resolution of the gas chromatography–mass spectrometry system [35].

In behavioral tests, bumblebees exhibited different behaviors in response to different volatiles. During evolution, bumblebees also developed a preference for floral chemical components related to their honey-gathering habits [36]. Olfactory signals are received by the antennae and then transmitted to the primary olfactory center, the antennal lobe [37,38]. The ovoid structure within the antennal lobe is the basic unit responsible for olfactory function. Different olfactory signals are processed through the spatiotemporal excitation patterns of specific ovoid structures [39] and then transmitted via nerves to the higher olfactory center, the mushroom body. Distinct characteristics of odors are processed separately by these nerve tracts [40,41]. In processing this olfactory signal, bumblebees are always interested in certain aromatic alcohols, but they show weaker responses or avoidance towards certain aromatic aldehydes, which is consistent with our research findings. Previous studies have shown that this may be due to these aldehydes not being commonly found in their environment, indicating that they are not food sources that lead to adaptive preferences for chemicals and highly specific olfactory receptors [7,42]. These physiological characteristics enable bumblebees to link ecological needs with behavioral habits, resulting in unique preferences for specific chemical components.

Insect behavior is associated with floral volatiles. Studies have shown that pollinators are attracted by floral color, size, olfactory cues, and nectar [43,44]. Many insects respond to either visual or olfactory stimuli alone [45–47], whereas bees rely on odor rather than visual cues for long-range orientation [45,48]. In the present study, we first collected and identified blueberry volatiles, and we then used EAG to rapidly screen compounds that elicited strong electrophysiological activity in *Bombus terrestris* and determined their optimal concentrations to narrow the scope of our behavioral assays. All our measurements were performed only using individual, not mixed, compounds. Given the large discrepancy between laboratory and field conditions, the active compounds should be blended at different ratios and systematically evaluated under greenhouse and open-field conditions for their effects on bumblebee visitation rate, pollination efficiency, and blueberry fruit set, so as to provide precise theoretical parameters and operational protocols for using *B. terrestris* in blueberry protected-culture pollination.

5. Conclusions

Floral fragrances play a crucial role in bumblebee pollination of plants. In this study, 6 classes totaling 32 volatile compounds were identified from blueberry flowers from Changbai Mountain. Here, we found that alcohols are the most abundant compounds identified. Different concentrations of these volatiles elicited clear electroantennographic (EAG) responses in Bombus terrestris. These bees showed strong EAG reactions to benzaldehyde, phenylpropyl aldehyde, citral, linalool, α-terpineol, and geraniol. Behavioral assays revealed that linalool, α-terpineol, and geraniol attracted the bumblebees, whereas phenylpropyl aldehyde repelled them, with higher concentrations of benzaldehyde and citral promoting bee avoidance.

Author Contributions: Conceptualization, D.L. and M.E.; Methodology, D.L.; Investigation and writing—review and editing, J.Z.; Formal analysis, Z.H.; Investigation, Z.H. and Y.C.; Resources, Y.C.; Data curation, Y.H. and M.E.; Writing—original draft, Y.H.; Writing—review and editing, K.H.M.; Project administration, M.E.; Funding acquisition, D.L. All authors have read and agreed to the published version of the manuscript.

Funding: This study was supported by the grant from the National Natural Science Foundation of China (Grant No.: 32272650) and the Natural Science Foundation of Changchun Normal University (Grant No.: 2017-008).

Institutional Review Board Statement: The study did not require ethical approval.

Informed Consent Statement: Not applicable.

Data Availability Statement: All data are available in this paper.

Acknowledgments: We thank Jiaru Zhang, Tianxu Qiu, Xiaoyan Zhu, Yuxian Zhou and Dian Zhang for their help with fieldwork.

Conflicts of Interest: The authors declare no conflicts of interest.

References

1. Zhao, X.; Jiang, J.; Pang, Z.; Ma, W.; Jiang, Y.; Fu, Y.; Liu, Y. Tracking existing factors directly affecting the reproduction of bumblebees: Current knowledge. *Insects* **2024**, *15*, 654. [CrossRef]
2. Zhang, H.; Huang, J.; Williams, P.H.; Vaissière, B.E.; Zhou, Z.; Gai, Q.; Dong, J.; An, J. Managed bumblebees outperform honeybees in increasing peach fruit set in China: Different limiting processes with different pollinators. *PLoS ONE* **2015**, *10*, e0121143. [CrossRef]
3. Symington, H.A.; Glover, B.J. Strawberry varieties differ in pollinator-relevant floral traits. *Ecol. Evol.* **2024**, *14*, e10914. [CrossRef]
4. Pfister, S.C.; Eckerter, P.W.; Schirmel, J.; Cresswell, J.E.; Entling, M.H. Sensitivity of commercial pumpkin yield to potential decline among different groups of pollinating bees. *R. Soc. Open Sci.* **2017**, *4*, 170102. [CrossRef]
5. Adebesin, F.; Widhalm, J.R.; Boachon, B.; Lefèvre, F.; Pierman, B.; Lynch, J.H.; Alam, I.; Junqueira, B.; Benke, R.; Ray, S.; et al. Emission of volatile organic compounds from petunia flowers is facilitated by an ABC transporter. *Science* **2017**, *356*, 1386–1388. [CrossRef]
6. Wakamura, S.; Arakaki, N.; Moriyama, D.; Kanayama, S.; Oike, M.; Kimura, A.; Wajima, S.; Ono, H.; Yasui, H. Does the orchid Luisia teres attract its male chafer pollinators (Scarabaeidae: Protaetia pryeri pryeri) by sexual deception? *Chemoecology* **2020**, *30*, 49–57. [CrossRef]
7. Schiestl, F.P. The evolution of floral scent and insect chemical communication. *Ecol. Lett.* **2010**, *13*, 643–656. [CrossRef]
8. Muhlemann, J.K.; Klempien, A.; Dudareva, N. Floral volatiles: From biosynthesis to function. *Plant Cell Environ.* **2014**, *37*, 1936–1949. [CrossRef]
9. Colazza, S.; Peri, E.; Cusumano, A. Chemical ecology of floral resources in conservation biological control. *Annu. Rev. Entomol.* **2023**, *68*, 13–29. [CrossRef]
10. Lyu, Z.; Zhou, T.; Sun, M.; Feng, M.; Guo, W.; Nie, L.; Song, Y.; Men, X.; Li, L.; Yu, Y. Exploratory comparison of flower visiting behavior and pollination ability of mason bees, bumble bees, and honey bees. *J. Econ. Entomol.* **2023**, *116*, 1949–1956. [CrossRef] [PubMed]
11. Switzer, C.M.; Combes, S.A.; Hopkins, R. Dispensing pollen via catapult: Explosive pollen release in mountain laurel (*Kalmia latifolia*). *Am. Nat.* **2018**, *191*, 767–776. [CrossRef]
12. Knauer, A.C.; Schiestl, F.P. Bees use honest floral signals as indicators of reward when visiting flowers. *Ecol. Lett.* **2015**, *18*, 135–143. [CrossRef]
13. Ferrão, L.F.V.; Sater, H.; Lyrene, P.; Amadeu, R.R.; Sims, C.A.; Tieman, D.M.; Munoz, P.R. Terpene volatiles mediates the chemical basis of blueberry aroma and consumer acceptability. *Food Res. Int.* **2022**, *158*, 111468. [CrossRef]
14. Sanchez-Ballesta, M.T.; Marti-Anders, C.; Álvarez, M.D.; Escribano, M.I.; Merodio, C.; Romero, I. Are the blueberries we buy good quality? comparative study of berries purchased from different outlets. *Foods* **2023**, *12*, 2621. [CrossRef]
15. Ma, L.; Sun, Z.; Zeng, Y.; Luo, M.; Yang, J. Molecular mechanism and health role of functional ingredients in blueberry for chronic disease in human beings. *Int. J. Mol. Sci.* **2018**, *19*, 2785. [CrossRef]
16. Cromie, J.; Ternest, J.J.; Komatz, A.P.; Adunola, P.M.; Azevedo, C.; Mallinger, R.E.; Muñoz, P.R. Genotypic variation in blueberry flower morphology and nectar reward content affects pollinator attraction in a diverse breeding population. *BMC Plant Biol.* **2024**, *24*, 814. [CrossRef]
17. Miñarro, M.; García, D. Complementary contribution of wild bumblebees and managed honeybee to the pollination niche of an introduced blueberry crop. *Insects* **2021**, *12*, 595. [CrossRef]
18. Bobiwash, K.; Uriel, Y.; Elle, E. Pollen foraging differences among three managed pollinators in the highbush blueberry (*Vaccinium corymbosum*) agroecosystem. *J. Econ. Entomol.* **2018**, *111*, 26–32. [CrossRef]
19. Estravis-Barcala, M.C.; Palottini, F.; Macri, I.; Nery, D.; Farina, W.M. Managed honeybees and South American bumblebees exhibit complementary foraging patterns in highbush blueberry. *Sci. Rep.* **2021**, *11*, 8187. [CrossRef]

20. Zhang, J.; Liu, J.; Gao, F.; Chen, M.; Jiang, Y.; Zhao, H.; Ma, W. Electrophysiological and behavioral responses of apis mellifera and bombusterrestris to melon flower volatiles. *Insects* **2022**, *13*, 973. [CrossRef]
21. Dötterl, S.; Vereecken, N.J. The chemical ecology and evolution of bee flower interactions: A review and perspectives. *Can. J. Zool.* **2010**, *88*, 668–697. [CrossRef]
22. Parachnowitsch, A.L.; Kessler, A. Pollinators exert natural selection on flower size and floral display in Penstemon digitalis. *New Phytol.* **2010**, *188*, 393–402. [CrossRef]
23. Wright, G.A.; Schiestl, F.P. The evolution of floral scent: The influence of olfactory learning by insect pollinators on the honestsignalling of floral rewards. *Funct. Ecol.* **2009**, *23*, 841–851. [CrossRef]
24. Howell, A.D.; Alarcón, R. Osmia bees (Hymenoptera: Megachilidae) can detect nectar-rewarding flowers using olfactory cues. *Anim. Behav.* **2007**, *74*, 199–205. [CrossRef]
25. Raman, V.; Wang, M.; Avula, B.; Lee, J.; Manfron, J.; Khan, I.A. Chemical mimicry in the corpse flower: Floral odor and phytochemical profiles of *Amorphophallus titanum* (becc.) becc. *Biochem. Syst. Ecol.* **2025**, *118*, 104920. [CrossRef]
26. Schiestl, F.P.; Johnson, S.D. Pollinator-mediated evolution of floral signals. *Trends Ecol. Evol.* **2013**, *28*, 307–315.
27. Majetic, C.J.; Raguso, R.A.; Ashman, T.L. The sweet smell of success: Floral scent affects pollinator attraction and seed fitness inHesperis matronalis. *Funct. Ecol.* **2009**, *23*, 480–487. [CrossRef]
28. Spaethe, J.; Brockmann, A.; Halbig, C.; Tautz, J. Size determines antennal sensitivity and behavioral threshold to odors in bumblebee workers. *Die Naturwissenschaften* **2007**, *94*, 733–739. [PubMed]
29. Yang, M.; Deng, G.C.; Gong, Y.B.; Huang, S.Q. Nectar yeasts enhance the interaction between Clematis akebioides and its bumblebee pollinator. *Plant Biol.* **2019**, *21*, 732–737. [CrossRef]
30. Liu, J.; Zhang, J.; Shen, J.; Zhao, H.; Ma, W.; Jiang, Y. Differences in EAG response and behavioral choices between honey bee and bumble bee to tomato flower volatiles. *Insects* **2022**, *13*, 987. [CrossRef]
31. Vendrame, S.; Daugherty, A.; Kristo, A.S.; Riso, P.; Klimis-Zacas, D. Wild blueberry (*Vaccinium angustifolium*) consumption improves inflammatory status in the obese zucker rat model of the metabolic syndrome. *J. Nutr. Biochem.* **2013**, *24*, 1508–1512. [CrossRef]
32. Rering, C.C.; Beck, J.J.; Hall, G.W.; McCartney, M.M.; Vannette, R.L. Nectar-inhabiting microorganisms influence nectar volatile composition and attractiveness to a generalist pollinator. *New Phytol.* **2017**, *220*, 750–759. [CrossRef]
33. Pichersky, E.; Gershenzon, J. The formation and function of plant volatiles: Perfumes for pollinator attraction and defense. *Curr. Opin. Plant Biol.* **2002**, *5*, 237–243. [CrossRef]
34. Rodriguez-Saona, C.; Parra, L.; Quiroz, A.; Isaacs, R. Variation in highbush blueberry floral volatile profiles as a function of pollination status, cultivar, time of day and flower part: Implications for flower visitation by bees. *Ann. Bot.* **2011**, *107*, 1377–1390. [CrossRef]
35. Silva, F.A.N.; da Silva, A.A.; de Sousa Fernandes, N.; Rodrigues, T.H.; Canuto, K.M.; do Nascimento, R.F.; de Brito, E.S.; deAragão, F.A.S.; Freitas, B.M.; Zocolo, G.J. Evaluation of Headspace Solid-Phase Microextraction Gas Chromatography-MassSpectrometry for the characterization of volatile organic compounds from Melon (*Cucumis melo* L.) Flowers. *Chromatographia* **2018**, *81*, 1231–1239. [CrossRef]
36. Huang, W.; Vallejo-Marín, M.; Inouye, D.W.; Yang, C.F.; Ye, Z.M. Bumblebees' flower preferences are associated with floral abundance and buzz frequency when buzz-pollinating co-flowering plants. *Entomol. Gen.* **2024**, *44*, 133–141. [CrossRef]
37. Kreiman, G.; MaBouDi, H.; Shimazaki, H.; Giurfa, M.; Chittka, L. Olfactory learning without the mushroom bodies: Spiking neural network models of the honeybee lateral antennal lobe tract reveal its capacities in odour memory tasks of varied complexities. *PLoS Comput. Biol.* **2017**, *13*, e1005551.
38. Lösel, P.D.; Monchanin, C.; Lebrun, R.; Jayme, A.; Relle, J.J.; Devaud, J.M.; Heuveline, V.; Lihoreau, M. Natural variability in bee brain size and symmetry revealed by micro-CT imaging and deep learning. *PLoS Comput. Biol.* **2023**, *19*, e1011529. [CrossRef]
39. Joerges, J.; Kuettner, A.; Galizia, C.G.; Menzel, R. Representations of odours and odour mixtures visualized in the honeybee brain. *Nature* **1997**, *387*, 285–287. [CrossRef]
40. Galizia, C.G.; Menzel, R. Odour perception in honeybees: Coding information in glomerular patterns. *Curr. Opin. Neurobiol.* **2000**, *10*, 504–510. [CrossRef]
41. Galizia, C.G.; Mcilwrath, S.L.; Menzel, R. A digital three-dimensional atlas of the honeybee antennal lobe based on optical sections acquired by confocal microscopy. *Cell Tissue Res.* **1999**, *295*, 383–394. [CrossRef]
42. Moreira, X.; Abdala-Roberts, L.; GolsBeatriz, R.; Lago-NúezSergio, B.; Rasmann, S.; Röder, G.; Soengas, P.; Vázquez-González, C.; Cartea, M.E. Insect herbivory but not plant pathogen infection drive floral volatile-mediated indirect effects on pollinators and plant fitness in Brassica rapa. *J. Ecol.* **2024**, *112*, 402–415. [CrossRef]
43. Clement, W.M. Flower color, a factor in attractiveness of alfalfa clones for honeybees. *Crop Sci.* **1965**, *5*, 267–268. [CrossRef]
44. Kauffeld, N.M.; Sorensen, E.L. *Interrelations of Honeybee Preference and Alfalfa Clones and Flower Color, Aroma, Nectar Volume and Sugar Concentration*; Agricultural Experiment Station, Kansas State University of Agriculture and Aplied Science: Manhattan, NY, USA, 1971; p. 14.

45. Dötterl, S.; Milchreit, K.; Schäffler, I. Behavioural plasticity and sex differences in host finding of a specialized bee species. *J. Comp.Physiol. A* **2011**, *197*, 1119–1126. [CrossRef]
46. Balkenius, A.; Rosen, W.; Kelber, A. The relative importance of olfaction and vision in a diurnal and a nocturnal hawkmoth. *J. Comp. Physiol. A* **2006**, *192*, 431–437. [CrossRef]
47. Wenzell, K.E.; Skogen, K.A.; Fant, J.B. Range-wide floral trait variation reflects shifts in pollinator assemblages, consistent with pollinator-mediated divergence despite generalized visitation. *Oikos* **2023**, *2023*, e09708. [CrossRef]
48. Dötterl, S.; Glück, U.; Jürgens, A.; Woodring, J.; Aas, G. Floral reward, advertisement and attractiveness to honey bees in dioecious Salix caprea. *PLoS ONE* **2014**, *9*, e93421. [CrossRef]

biology

MDPI

Article

The Thermal Adaptability of *Sclerodermus guani* Xiao *et* Wu (Hymenoptera: Bethylidae), an Important Parasitoid of Long-Horned Beetles in China

Lina Wang [1], Yuhua Situ [2], Jie Zhang [1], Kui Kang [1], Zhongjiu Xiao [1], Shaobo Wang [1], Ke Wei [3,*] and Yanlong Tang [1,*]

1 College of Biology and Agriculture, Zunyi Normal University, Zunyi 563002, China
2 Centre of Earth Observation Science, University of Manitoba, Winnipeg, MB R3T 2N2, Canada
3 Key Laboratory of Forest Protection of National Forestry and Grassland Administration, Ecology and Nature Conservation Institute, Chinese Academy of Forestry, Beijing 100091, China
* Correspondence: weike@caf.ac.cn (K.W.); 15120086160@163.com (Y.T.)

Simple Summary: *Sclerodermus guani* Xiao *et* Wu is an important parasitoid of long-horned beetles in China and has been widely used to control forest trunk-boring pests for over 40 years. This study aimed to determine the adaptive characteristics of this parasitoid wasp in response to temperature variations. Between 21 and 33 °C, the parasitoid behavior and parasitism capacity of the wasp varied significantly. Higher temperatures resulted in more-active females, a shorter supplementary feeding period, and a reduced developmental duration in offspring. At the highest temperature (33 °C), offspring development was most rapid, but the number of offspring was the lowest. Conversely, at the lowest temperature (21 °C), offspring development was the slowest, the number of offspring was similarly low, and parasitism and offspring emergence rates were at their minimum. These results indicate that *S. guani* exhibits higher developmental fitness between 24 and 30 °C, while its reproductive fitness is the highest between 27 and 30 °C. Therefore, it is recommended that this parasitoid wasp be reared indoors at 27 and 30 °C and released for pest control in forests when the temperature is above 24 °C.

Abstract: *Sclerodermus guani* is the most extensively studied and widely used parasitoid wasp for controlling forest trunk-boring pests in China, with temperature being a key environmental factor affecting its development and reproduction. Against the backdrop of global climate change, the frequency of heat stress on parasitoid wasps is increasing. Therefore, studying their adaptability to heat stress is of great significance for breeding heat-tolerant strains. The findings of such research may help improve efficiency in rearing this wasp and provide guidance for its use in pest control. This study investigated the thermal adaptability of the wasp at temperatures ranging from 21 to 33 °C. The results showed that temperature significantly affected its parasitism capacity and behavior. As temperature increased, female wasps searched for and stung hosts more actively, and their pre-oviposition period decreased, as well as their developmental duration. The reproductive fitness of the wasp also varied significantly with temperature, with the most favorable temperatures being 27 °C and 30 °C. Both low and high temperatures were detrimental to its reproduction, and the difference in the number of female offspring was significant. The experimental results indicate that *S. guani* has higher reproductive fitness at 27–30 °C. Therefore, it is recommended that this parasitoid wasp be reared indoors at 27–30 °C and released for pest control in forests when the temperature is above 24 °C.

Keywords: *Sclerodermus guani* Xiao *et* Wu; temperature; development; reproduction

1. Introduction

The impacts of wood-boring beetles on forests and crops necessitate more-efficient parasitoids under climate change scenarios. *Sclerodermus guani* was discovered in the 1970s in China's Shandong and Guangdong provinces [1,2], and it was subsequently found to parasitize the larvae and pupae of small- to medium-sized long-horned beetles, such as *Saperda populnea* and *S. sinoauster* [3]. It can also effectively parasitize the first- to third-instar larvae of larger long-horned beetles such as *Monochamus alternatus* and *Anoplophora glabripennis* in China [4]. Consequently, *S. guani* has gradually become the most intensively studied and widely used parasitoid wasp for controlling forest trunk-boring pests in China [4]. In this country, the life stages at which these beetle pests are most susceptible to parasitism by *S. guani* largely occur in July and August. For example, in this period, *S. bifasciatus* and *S. sinoauster* are in their mature larval stage, and while *M. alternatus* and *A. glabripennis* are mostly in their first to third larval instars [4]. July and August are also the hottest months in the Northern Hemisphere. In recent years, as global warming has made high summer temperatures the norm, studying the thermal adaptation characteristics of this parasitoid wasp, which inhabits the warm, temperate, monsoon climate zone of the Northern Hemisphere, has become of great significance for its mass rearing and application.

Ambient temperature is a primary determinant of an insect's performance and ecological success [5,6]. Changes in temperature can affect an organism's specific biological processes, including behavior, growth and developmental rates, fecundity, and physiological metabolism [7–9]. Generally, the developmental duration of parasitoid wasps shortens as temperature increases, as seen in *Tetrastichus planipennisi* [10], *Nasonia vitripennis* [11], *Cyanopterus ninghais* [12], *S. alternatusi* [13], and *S. pupariae* [14]. However, at higher temperatures, the reproduction of parasitoid wasps is limited. For example, while no parasitoid eggs of *T. planipennisi* hatch at 35 °C [10], the eggs of *N. vitripennis* also fail to hatch at 34 °C [11], and the number of offspring of *S. alternatusi* and *S. pupariae* decreases significantly at 33 °C [13,14]. Temperature also has a significant impact on the parasitoid behavior and activity of wasps. The parasitism rates of *S. alternatusi* and *S. pupariae* decline markedly at 33 °C [13,14], whereas the parasitism rate of *C. ninghais* is the highest at 29 °C, with both low and high temperatures adversely affecting its parasitoid efficiency [12]. Understanding how *S. guani* tolerates high temperatures is essential for its use in biocontrol under climate change scenarios, yet no data exist for temperatures > 30 °C. In addition, under rising temperature conditions, the developmental duration of the host is shortened, with environmental temperature variations reducing the host presence window, thus shortening the period during which hosts at susceptible stages are available [15].

Wasps of the genus *Sclerodermus* are ectoparasitoids. Before oviposition, they undergo a series of processes including searching for a host, stinging it and injecting it with venom to paralyze it, cleaning debris from the surface of the host's body, and feeding on the host's hemolymph for nutritional supplementation [16,17]. Temperature significantly affects the vitality of these wasps. For *S. alternatusi*, the time from introduction to host paralysis is 6.7 days at 21 °C but only 1.5 days at 30 °C; the subsequent period until oviposition is 9.3 days long at 21 °C but only 4.4 days long at 33 °C [13]. For *S. pupariae*, the time from introduction to oviposition is about 12 days at 21 °C, shortening to less than 3 days at 33 °C [14]. In the case of *S. sichuanensis*, the supplementary feeding period is about 20 days in April (spring) but only about 6 days in July (summer) [18]. However, the relationship between the supplementary feeding period of *S. guani* and temperature has not been reported. Although a few studies have described the effects of temperature on the development and reproduction of *S. guani*, none have included data collected at high temperatures above 30 °C [3,19]. Against the backdrop of global warming, insects are experiencing heat stress more frequently, and parasitoid wasps are no exception. The response and adaptability of parasitoid wasps to high temperatures are therefore current research hotspots. Here, we examined the effects of high temperatures (>30 °C) on the development, reproduction, and parasitism efficiency of *S. guani*, with the aim of elucidating its thermal tolerance and potential in biocontrol under climate change.

2. Materials and Methods

2.1. Insects

Sclerodermus guani was collected in 2022 at the Xishan Experimental Forest Farm in Beijing (116.19° E, 39.99° N) and reared on larvae of *Thyestilla gebleri*, which acted as hosts, in the laboratory. The rearing conditions were a temperature of 25 °C, 60–70% relative humidity (RH), a photoperiod of 8 h L:16 h D, and a light intensity of 3000 lx. Healthy and active 3-day-old female wasps that had been fully mated were selected for this study. The host insects used were last-instar larvae of *Thyestilla gebleri* weighing between 200.0 and 220.0 mg, with thirty replicates for each treatment collected from the roots of infested *Abutilon theophrasti* Medicus from Dagang District (38°56′ N, 117°29′ E), Tianjin City, China [20,21].

2.2. Inoculation of *Sclerodermus guani*

Individual hosts were placed in glass culture tubes (1.0 cm in diameter, 5.0 cm in length). Female wasps of similar size, with a body length of about 3.5 mm, were selected and introduced individually into a tube containing one *T. gebleri* larva. The tube openings were then sealed with cotton wool plugs, and water/sugar was not provided. The rearing temperatures were set to 21 °C, 24 °C, 27 °C, 30 °C, and 33 °C, with 30 tubes prepared for each treatment. Across temperature conditions, the humidity and photoperiod were kept constant (RH = 60–70%; photoperiod of 8 h L:16 h D; light intensity of 3000 lx).

2.3. Observation of the Development and Reproductive Fitness of *Sclerodermus guani* Under Different Rearing Temperatures

The parasitism capacity of the mother wasp on the host was recorded, including the parasitism rate and the offspring emergence rate. The parasitism rate was defined as the proportion of parasitized hosts to the total number of hosts provided. The offspring emergence rate was defined as the proportion of hosts from which offspring wasps emerged to the total number of hosts provided. The pre-oviposition period of the mother wasp, and the developmental duration and number of offspring wasps were also recorded. The pre-oviposition period is the interval from the introduction of the wasp to the observation of the first laid egg with a Motic stereoscope. The developmental duration of the offspring wasps included the egg, larval, pupal, and egg-to-adult durations. The egg duration is the interval from the laying of the first egg to the hatching of the first larva. The larval duration is the interval from the hatching of the first larva to the appearance of the first cocoon. The pupal duration is the interval from the appearance of the first cocoon to the emergence of the first adult wasp. Finally, the egg-to-adult duration is the time from the laying of the first egg to the emergence of the first adult wasp. Duration was measured in days. For each culture tube, the developmental duration of the offspring wasps was recorded once to represent the developmental progress of the offspring cohort, including the times at which the first egg was laid by the female, the first egg hatched, a first-instar larva cocooned, and the first adult emerged. After the complete emergence of the offspring wasps, the number of female and male wasps from each host and the sex ratio were calculated. The sex ratio is defined as the proportion of female wasps in all offspring of a single host [21,22]. All the above measurements were made once daily.

2.4. Developmental Threshold Temperature and Effective Accumulated Temperature for Each Life Stage of Sclerodermus guani

Based on the linear relationship between temperature and developmental rate, the effective accumulated temperature (K) and the developmental threshold temperature (C) required for *S. guani* to complete a specific life stage were calculated. The developmental rate (V) at different temperatures for each life stage was calculated using the following Equation:

$$V = \frac{1}{N}\left(d^{-1}\right) \quad (1)$$

where V is the daily developmental rate, and N is the number of days required for completion of each life stage of *S. guani*, obtained from the experiment in Section 2.3. A linear regression Equation,

$$T = a + b \times V \quad (2)$$

was fitted with the experimental temperature (T) as the y-axis and the developmental rate (V) as the x-axis. Following standard linear methods [3], the x-intercept of the regression corresponds to the developmental threshold temperature (C), and the inverse of the slope represents the effective accumulated temperature (K).

2.5. Data Analysis and Statistics

Fisher's exact test was used to examine the differences in the parasitism capacity of *S. guani* under different temperatures. For the pre-oviposition period, developmental duration, number of offspring, and sex ratio, data were first assessed for normality using the Shapiro–Wilk test and for homogeneity of variance using Levene's test. Female sex ratio data were arcsine-transformed to meet these assumptions. One-way ANOVA was then performed to test for significant differences among treatments, followed by Fisher's least significant difference (LSD) test for multiple comparisons. The linear regression equation was evaluated using an F-test. All statistical analyses were performed using SPSS 22.0.

3. Results

3.1. Effect of Temperature on the Parasitism Capacity of Sclerodermus guani

The parasitism capacity of *S. guani* was significantly influenced by temperature (Table 1). Across the five temperature gradients tested (21 °C to 33 °C), both the parasitism rate ($\chi^2 = 11.606$; $df = 4$; $p = 0.021$) and the F1 emergence rate ($\chi^2 = 29.667$; $df = 4$; $p < 0.001$) were significantly different. At the lowest temperature of 21 °C, the parasitism and emergence rates were at their minimum, at 66.67% and 46.67%, respectively. Both rates peaked at 96.67% at 24 °C. Subsequently, as temperatures increased further, both the parasitism and emergence rates exhibited a declining trend, with 90.00% at 27 °C, 86.67% at 30 °C, and a further decrease to 83.33% at 33 °C.

Table 1. Parasitism rate and emergence rate of *Sclerodermus guani* reared at different temperatures.

Temperature/°C	Parasitized Host/n	Parasitism Rate/%	Number of Hosts with Parasitoid Offspring/n	Emergence Rate/%
21	20	66.67 b	14	46.67 b
24	29	96.67 a	29	96.67 a
27	27	90.00 ab	27	90.00 a

Table 1. *Cont.*

Temperature/°C	Parasitized Host/n	Parasitism Rate/%	Number of Hosts with Parasitoid Offspring/n	Emergence Rate/%
30	26	86.67 ab	26	86.67 a
33	25	83.33 ab	25	83.33 a
χ^2	/	11.606	/	29.667
p	/	0.021	/	<0.001

Note: Data are means ± SE of thirty replicates. Values with different lowercase letters in the same column are significantly different at the 0.05 level according to Fisher's exact test.

3.2. The Effect of Temperature on the Parasitoid Behavior of Sclerodermus guani

The results demonstrated that temperature significantly affects the parasitoid behavior of *S. guani*. Specifically, higher temperatures increased the activity level of the parasitoid wasps, leading to more-proactive host-seeking and stinging behaviors. Consequently, the pre-oviposition period was significantly shortened ($F = 795.822$; $df = 4, 126$; $p < 0.001$) (Figure 1). A significant negative correlation was observed between the pre-oviposition period and temperature ($y = -16.24\ln(x) + 25.783$). At 21 °C, the pre-oviposition period was longest, averaging 31.3 days. This duration shortened dramatically to an average of 6.0 days at 24° C and continued to decrease to 5.3 and 4.5 days at 27 °C and 30 °C, respectively. The shortest pre-oviposition period was recorded at 33 °C, at which the wasps required only 4.1 days to complete host paralysis and nutritional feeding before commencing oviposition (Table S1).

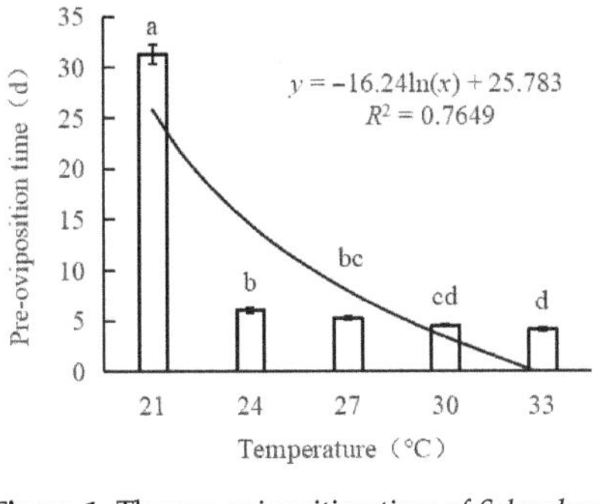

Figure 1. The pre-oviposition time of *Sclerodermus guani* females at different temperatures. Data are means ± SE of thirty replicates. Different letters above the columns indicate significant differences among the groups according to ANOVA performed with the LSD test ($p < 0.05$).

3.3. Developmental Adaptability of Sclerodermus guani to Different Temperatures

Sclerodermus guani exhibited distinct developmental adaptations to varying temperatures. The durations of the egg stage ($F = 20.769$; $df = 2, 82$; $p < 0.001$), larval stage ($F = 23.239$; $df = 2, 78$; $p < 0.001$), pupal stage ($F = 74.504$; $df = 2, 77$; $p < 0.001$), and the complete egg-to-adult stage ($F = 151.390$; $df = 2, 77$; $p < 0.001$) all decreased significantly with rising temperatures (Figure 2). The longest developmental periods were recorded at 21 °C, with average egg, larval, pupal, and egg-to-adult durations of 7.5, 11.3, 41.1, and 59.3 days, respectively. At 24 °C, these durations shortened significantly to 6.1, 6.4, 15.3, and 27.8 days, respectively. The shortest developmental times occurred at 33 °C, with durations of 4.6, 4.8, 11.1, and 20.4 days for the respective stages. Compared to the durations at 21 °C, these represent reductions of 2.9, 6.5, 30.0, and 38.9 days. The durations of the egg, larval, pupal, and egg-to-adult stages were all negatively correlated with temperature, as described by the following respective equations: $y = -2.451\ln(x) + 7.1511$; $y = -3.381\ln(x) + 10.597$; $y = -17.59\ln(x) + 35.925$; and $y = -23.12\ln(x) + 53.294$ (Table S1).

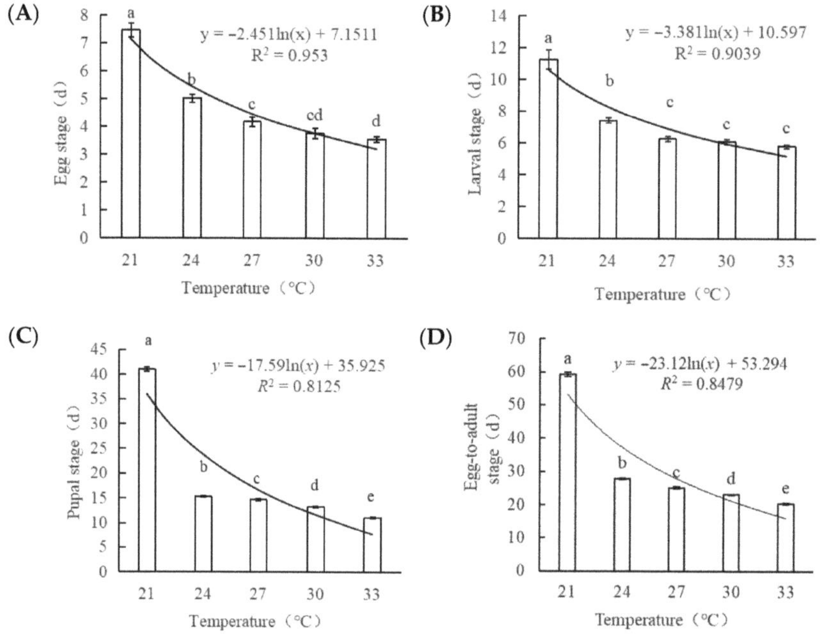

Figure 2. The developmental stages of *Sclerodermus guani* at different temperatures. (**A**): egg stage; (**B**): larval stage; (**C**): pupal stage; (**D**): egg-to-adult stage. Data are means ± SE of thirty replicates. Different letters above the columns indicate significant differences among the groups according to ANOVA performed with the LSD test ($p < 0.05$).

3.4. Reproductive Fitness of Sclerodermus guani at Different Temperatures

Sclerodermus guani exhibited significant differences in reproductive fitness across various temperatures. The optimal temperatures for reproduction were 27 °C and 30 °C, which yielded the most total progeny, averaging 43.0 and 42.5 individuals/tube, respectively (Figure 3). Conversely, both lower and higher temperatures were detrimental to reproduction. At 21 °C, the total offspring count was 31.4 per tube, while at 33 °C, it was at its lowest, at only 28.8 per tube—a significant reduction of 13.7 individuals from the peak value ($F = 11.154$; $df = 4, 120$; $p < 0.001$). The number of male offspring did not differ significantly across temperatures ($F = 0.72$; $df = 4, 120$; $p = 0.580$), with averages ranging from 2.2 to 2.9 individuals/tube. In contrast, the number of female offspring varied significantly ($F = 12.598$; $df = 4, 120$; $p < 0.001$). The most female progeny were observed at 27 °C (mean of 40.1 per tube), followed closely by 30 °C (mean of 39.9 per tube). However, the number dropped sharply to 26.6 per tube at 33 °C, 13.5 individuals fewer than the maximum (Table S2). Despite these variations, the sex ratio was consistently female-biased, with the proportion of females exceeding 90% and showing no significant differences across all treatments ($F = 0.615$; $df = 4, 120$; $p = 0.652$) (Table S3). This confirms that *Sclerodermus guani* is a species with a typically female-biased sex ratio.

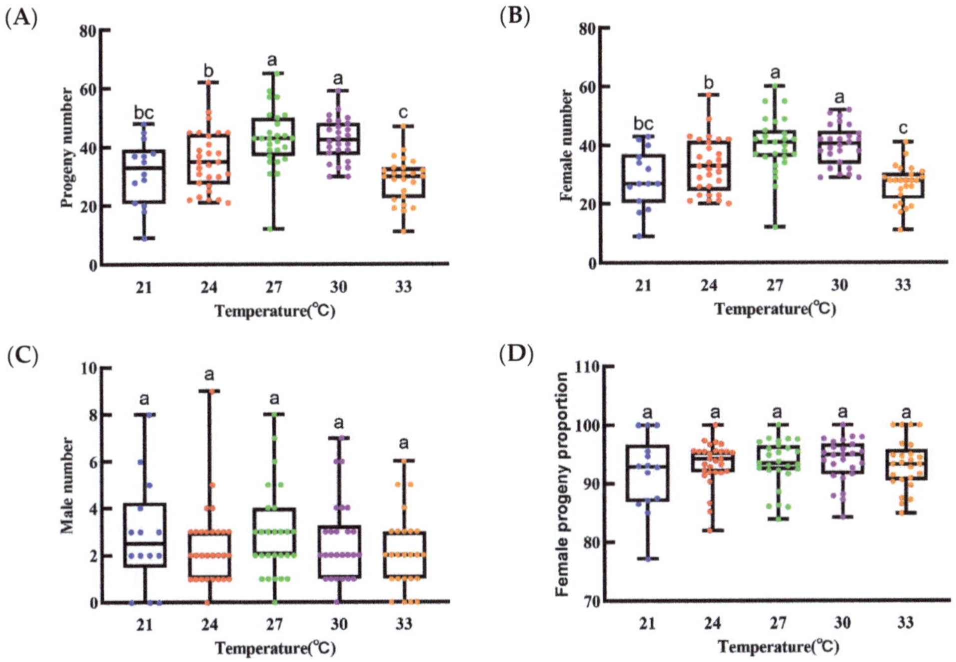

Figure 3. The brood size and female ratio of *Sclerodermus guani* progeny. (**A**): Progeny number; (**B**): Female number; (**C**): male number; (**D**): female ratio. Data are means ± SE of thirty replicates. Different letters above the bars indicate significant differences among the groups according to ANOVA performed with the LSD test ($p < 0.05$).

3.5. Developmental Thresholds and Developmental Threshold Temperatures for Different Life Stages of Sclerodermus guani

The developmental rates of *S. guani* at various temperatures are presented in Table 2. The rates for the egg, larval, and pupal stages ranged from 0.1368 to 0.2867, 0.0932 to 0.1741, and 0.0244 to 0.0908, respectively. Based on these data, the lower developmental threshold temperature (T0) for the egg stage was estimated to be 10.19 °C, with an effective accumulated temperature (K) of 72.57 degree-days. For the larval stage, the developmental threshold was 7.73 °C and the effective accumulated temperature was 131.21 degree-days. Finally, for the pupal stage, the developmental threshold was 15.57 °C and the effective accumulated temperature was 176.02 degree-days (Table 3).

Table 2. The developmental rates of *Sclerodermus guani* among different groups.

Temperature (°C)	Developmental Rates		
	Egg Stage	Larval Stage	Pupal Stage
21	0.1368	0.0932	0.0244
24	0.2034	0.1365	0.0653
27	0.2506	0.1626	0.0684
30	0.2807	0.1678	0.0758
33	0.2867	0.1741	0.0908

Table 3. The lower developmental threshold temperature and effective accumulated temperature of *Sclerodermus guani*.

Stage	Lower Developmental Threshold Temperature (°C)	Effective Accumulated Temperature (Degree-Days)	Model	R^2	F	p
Egg stage	10.19	72.57	$y = 72.57x + 10.19$	0.912	31.148	0.011
Larval stage	7.73	131.21	$y = 131.21x + 7.73$	0.845	16.302	0.027
Pupal stage	15.57	176.02	$y = 176.02x + 15.57$	0.841	15.846	0.028

4. Discussion

Currently, all reported parasitoids of the genus *Sclerodermus* are synovigenic. Females rely on host-feeding to supplement their nutrition for the development of their ovaries and oocytes, whereas males can mate without nutrient supplementation [1,18,23,24]. After female wasps supplement their nutrition, their abdominal intersegmental membranes stretch, and their body length increases by more than one-third [25]. Studies have shown that the duration of the nutrient supplementation period for females of the genus *Sclerodermus* is closely related to temperature. For *S. alternatusi*, this period can be as long as 15 days at 21 °C, but it shortens rapidly when the temperature rises above 24 °C [13]. For *S. guani*, the pre-oviposition period is as long as 31.3 days at 21 °C, shortening significantly at 24 °C, and at 33 °C, the processes of stinging and paralyzing the host and nutrient supplementation only take 4.1 days before oviposition begins. Observations have found that at low temperatures, most *S. guani* wasps are inactive and not proactive in searching for and stinging hosts. However, as the temperature increases, their parasitoid behavior changes significantly. This suggests that releasing *S. guani* wasps for pest control when the temperature is above 24 °C can significantly improve their parasitism efficiency.

Within a suitable temperature range, the developmental rate generally accelerates as the temperature rises, but the temperature range for maximum fecundity is often narrower [13,14]. This study found that under five temperature conditions from 21 °C to 33 °C, the developmental rates for the egg, larval, and pupal stages of *S. guani* all increased with rising temperature, with their corresponding developmental durations shortening. At 33 °C, the egg, larval, pupal, and egg-to-adult durations were shorter by 2.9 d, 6.5 d, 30 d, and 38.9 d, respectively, compared to those at 21 °C. However, the optimal temperature range for the reproduction of *S. guani* was only between 27 °C and 30 °C, with both low and high temperatures being unfavorable in this regard. At 21 °C, the total number of offspring was 31.4 individuals/tube, which was 11.6 individuals/tube fewer than the maximum. The reduction of 13.7 offspring at 33 °C represents a 30% decrease in reproductive output, which could significantly affect mass-rearing efficiency.

Under rising temperature conditions, the developmental duration of parasitoids at immature stages does not invariably shorten. For example, the developmental duration of *Venturia canescens* gradually shortens as the temperature increases from 15 °C to 30 °C, but it becomes prolonged when the temperature exceeds 30 °C [26]. Similarly, the developmental duration of *Zele chlorophthalmus* shortens with increasing temperature, but it becomes longer at 29 °C than at 25 °C [27]. In the temperature range studied here, the phenomenon of a prolonged developmental duration with increasing temperature was not observed for *S. guani*. This may be because the set temperatures did not reach the species' non-optimal developmental threshold; that is, 33 °C is still a suitable developmental temperature for *S. guani*. This result is consistent with a previous study, which found that temperatures needed to rise to 35 °C to be unfavorable for the parasitism and development of *S. sichuanensis* [28].

Parasitoids reared under constant laboratory temperatures for long periods may face greater challenges when confronted with variable temperature conditions in the field. Particularly in the context of current global climate warming, parasitoids are experiencing increased frequency of heat stress. Research indicates that parasitoids have a unique nervous system that can perceive changes in the external environment and adapt by adjusting physiological and biochemical processes [29,30]. When facing heat stress, parasitoids can utilize protective enzymes such as Catalase (CAT), Peroxidase (POD), and Superoxide dismutase (SOD) to maintain the dynamic equilibrium of oxidative metabolism in their bodies [31–33]. High temperatures can lead to protein degradation in insects, thereby affecting their cellular functions [34]. Studies show that the genes responsible for thermotolerance in insects are primarily concentrated in the heat shock protein (HSP) family [35]. Generally, the expression level of HSPs in insects is positively correlated with their thermotolerance [36,37]. Research has shown that the expression of the heat shock protein *Tchsp70* in *Trichogramma chilonis Isshii* increased by 7.41-fold and 13.47-fold at 32 °C and 40 °C, respectively, compared to at a control temperature (25 °C) [38]. In our preliminary research, we have already detected heat shock protein genes in wasps of the genus *Sclerodermus*. Our team will subsequently conduct systematic research on their functions to clarify the adaptive mechanisms of *Sclerodermus* in response to heat stress, providing a basis for breeding heat-tolerant strains of parasitoids.

Studies have shown that *Sclerodermus* parasitoids are of a completely inbreeding type; males mate with their female siblings from the same brood immediately after eclosion [39]. This may be associated with the low population density of these parasitoids in the wild and the difficulty in finding non-sibling mates, as they live within the boring tunnels of wood-boring pests. To ensure the presence of male offspring in the subsequent generation, female wasps of the genus *Sclerodermus* will first lay unfertilized eggs, which develop into males. However, an excess of males is detrimental to population development [40]. Therefore, the natural sex ratio of many bethylid wasps is highly stable and female-biased [16,39]. Research indicates that factors such as the number of maternal females, host size, and temperature all affect the offspring sex ratio [41,42]. This study found that temperature has a minor effect on the female sex ratio of *S. guani*; across all five temperatures, the female ratio was greater than 90%, demonstrating the stability of this biological characteristic in *S. guani*.

Parasitoids of the genus *Sclerodermus*, represented by *S. guani*, *S. sichuanensis*, *S. pupariae*, and *S. alternatusi*, are important natural enemy insects widely used in Chinese forestry. Billions of these wasps need to be released annually to control pests [4,43,44]. Inundative release of natural enemies requires the support of robust factory-based mass-rearing technology [45]. An increase in temperature is clearly beneficial in shortening the reproductive cycle of *Sclerodermus* wasps. However, at high temperatures, not only does the parasitism rate of the wasps decrease, but the factitious hosts are also more prone to becoming moldy and turning black, and the number of offspring significantly declines. In summary, we believe that rearing temperatures that are too high or too low are detrimental to the reproductive fitness of *S. guani*. According to the fecundity and parasitism rate of the wasp, the optimal temperature range for the reproduction of *S. guani* is only between 27 °C and 30 °C, and the optimal temperature at which they may be released in the field for pest control is above 24 °C.

5. Conclusions

Temperature significantly affects the parasitism capacity and behavior of *Sclerodermus guani*. At low (21 °C) and high (33 °C) temperatures, the parasitism rate is lower than at

24–30 °C. The higher the temperature, the more actively the female wasp seeks out and stings the host, and the shorter its pre-oviposition period.

The developmental duration of *S. guani* shortens as temperature increases. Both high (33 °C) and low (21 °C) temperatures are detrimental to the reproduction of *S. guani*. The observed reduction of 13.7 offspring at 33 °C represents a 30% decrease in reproductive output.

When temperatures exceed 30 °C, although the parasitism rate remains above 80%, the number of eggs laid per female decreases to 60% of that at a normal temperature. This will directly affect the economic efficiency of mass rearing and the parasitism effectiveness in the field. We recommend prioritizing the release of wasps during the 27–30 °C period in high-temperature regions, along with implementing short-term cold storage and transportation protocols.

Supplementary Materials: The following supporting information can be downloaded at: https://www.mdpi.com/article/10.3390/biology14091234/s1, Table S1: Difference in Duration (days) is a record of the duration of each stage; Table S2: Offspring Count shows the number of male and female offsprings and the sex ratio; Table S3: Female Sex Ratio shows the Female Sex Ratio and Female Sex Ratio after Natural Logarithm transformation.

Author Contributions: Conceptualization, Y.S., K.W., and Y.T.; methodology, Y.S., and Y.T.; software, Y.S. and J.Z.; validation, J.Z., L.W. and K.W.; formal analysis, K.K. and S.W.; investigation, Y.S., J.Z. and L.W.; resources, S.W. and K.W.; data curation, J.Z., S.W. and Z.X.; writing—original draft preparation, Y.S. and Y.T.; writing—review and editing, Y.T. and K.W.; visualization, Y.T.; supervision, Y.T.; project administration, Y.T.; funding acquisition, K.K. and Z.X. All authors have read and agreed to the published version of the manuscript.

Funding: This research was funded by the Natural Science Research Project of Guizhou Provincial Department of Education, grant number QJJ[2024]343; Fundamental Research Funds of the Chinese Academy of Forestry, CAFYBB2024MA017; the Guizhou Provincial Department of Education Innovation Team Project, grant number KY[2020]026; and the Science Foundation of Guizhou Science and Technology Department, QKHZK[2023]480.

Institutional Review Board Statement: Not applicable.

Informed Consent Statement: Not applicable.

Data Availability Statement: The original contributions presented in this study are included in the article/Supplementary Materials. Further inquiries can be directed to the corresponding authors.

Conflicts of Interest: The authors declare no conflicts of interest.

Abbreviations

The following abbreviations are used in this manuscript:

RH	relative humidity
ANOVA	analysis of variance
LSD	least significant difference
SE	standard error
CAT	catalase
POD	peroxidase
SOD	superoxide dismutase
HSPs	heat shock proteins

References

1. Xiao, G.; Wu, J. A new Species of *Sclerodermus* from China (Hymenoptera, bethylidae). *Sci. Silvae Sin.* **1983**, *19*, 81–84. (In Chinese)
2. Yang, Z.; Wang, X.; Cao, L.; Yao, Y.; Tang, Y. Re-description of *Sclerodermus guani* and Revision of the Genus (Hymenoptera:Bethylidae) in China. *Chin. J. Biol. Control.* **2014**, *30*, 1–12. (In Chinese)

3. Yao, D.; Yan, J.; Xu, C.; Qin, X.; Gao, R.; Li, Y. Studies on the Influence of Temperature and Humidity on the Development of Bethylid parasitoid Wasp (*Sclerodermus guani* xiao *et* Wu). *Sci. Silvae Sin.* **1983**, *19*, 85–91. (In Chinese)
4. Yang, Z.; Wang, X.; Zhang, Y. Recent Advances in Biological Control of Important Native and Invasive Forest Pests in China. *Biol. Control.* **2014**, *68*, 117–128. [CrossRef]
5. Angilletta, M. *Thermal Adaptation: A Theoretical and Empirical Synthesis*; OUP Oxford ProQuest Ebook Central: Oxford, UK, 2009.
6. Régnière, J.; Powell, J. Animal Life Cycle Models (Poikilotherms). In *Phenology: An Integrative Environmental Science*, 2nd ed.; Schwartz, M.D., Ed.; Springer: Dordrecht, The Netherlands, 2013; pp. 295–316.
7. Franke, K.; Fischer, K. Effects of Inbreeding and Temperature Stress on Life History and Immune Function in a Butterfly. *J. Evolution. Biol.* **2013**, *26*, 517–528. [CrossRef] [PubMed]
8. Bauerfeind, S.; Fischer, K. Integrating Temperature and Nutrition-environmental Impacts on an Insect Immune System. *J. Insect Physiol.* **2014**, *64*, 14–20. [CrossRef]
9. Moore, M.; Kester, K.; Kingsolver, J. Rearing Temperature and Parasitoid Load Determine Host and Parasitoid Performance in *Manduca sexta* and *Cotesia congregata*. *Ecol. Entomol.* **2020**, *45*, 79–89.
10. Duan, J.; Schmude, J.; Wang, X.; Watt, T.; Bauer, L. Host Utilization, Reproductive Biology, and Development of the Larval Parasitoid *Tetrastichus planipennisi* as Influenced by Temperature:Implications for Biological Control of the Emerald Ash Borer in North America. *Biol. Control.* **2018**, *125*, 50–56. [CrossRef]
11. Zhang, Y.; Wang, Y.; Liu, C.; Wang, J.; Hu, G.; Wang, M.; Yang, L.; Chu, J. Development of *Nasonia vitripennis* (Hymenoptera: Pteromalidae) at Constant Temperatures in China. *J. Med. Entomol.* **2019**, *56*, 368–377. [CrossRef]
12. Wang, S.; Han, M.; Wei, K.; Wang, X. Effects of Temperature on Reproduction and Development of *Cyanopterus ninghais* (Hymenoptera: Braconidae), a Larval Parasitoid of *Monochamus alternatus* (Coleoptera: Cerambycidae). *J. Appl. Entomol.* **2024**, *148*, 415–423. [CrossRef]
13. Chen, R.; Tang, Y.; Tang, H.; Wang, X.; Yang, Z. Effect of Temperature on Development and Reproduction of *Sclerodermus alternatus*. *For. Res.* **2019**, *32*, 114–119. (In Chinese)
14. Hu, S.; Wang, X.; Yang, Z.; Chen, R. Optimal Temperatures for Artificial Rearing of Parasitoid, *Sclerodermus pupariae* (Hymenoptera: Bethylidae). *Chin. J. Biol. Control.* **2019**, *35*, 343–349. (In Chinese)
15. Carmelo, P.B.; Gregorio, V.; Umberto, B. Environmental Thermal Levels Affect the Phenological Relationships between the Chestnut Ggall Wasp and its Parasitoids. *Physiol. Entomol.* **2019**, *44*, 87–98.
16. Tang, X.; Li, M.; Kapranas, A.; Xu, F.; Hardy, L.C.; Li, B. Mutually Beneficial Host Exploitation and Ultra-biased Sex Ratios in Quasisocial Parasitoids. *Nat. Commun.* **2014**, *5*, 4942. [CrossRef] [PubMed]
17. Gao, S.; Wei, K.; Tang, Y.; Wang, X.; Yang, Z. Effect of Parasitoid Density on the Timing of Parasitism and Development Duration of Progeny in *Sclerodermus pupariae* (Hymenoptera: Bethylidae). *Biol. control.* **2016**, *97*, 57–62. [CrossRef]
18. Zhou, Z.; Yang, W.; Zeng, C.; Yang, D.; Ye, W. A Preliminary Study on the Bionomics of *Sclerodermus sichuanensis* (Hymenoptera: Bethylidae). *Sci. Silvae Sin.* **1997**, *33*, 475–480. (In Chinese)
19. Wang, J.; Zhang, Q.; Luo, J.; Chen, B. Combined Interruption Effect of Temperature and Parasitoid Density on Parasitoids of *Sclerodermus guani*. *SW Chin. J. Agri. Sci.* **2010**, *23*, 243–246. (In Chinese)
20. Wei, K.; Tang, Y.; Wang, X.; Cao, L.; Yang, Z. The Developmental Strategies and Related Profitability of an Idiobiont Ectoparasitoid *Sclerodermus pupariae* Vary with Host Size. *Ecol. Entomol.* **2014**, *39*, 101–108. [CrossRef]
21. Kang, K.; Wang, L.; Xiao, Z.; Wang, S.; Wei, K.; Wang, X.; Zhang, Y.; Tang, Y. Effects of Photoperiod on the Developmental Duration and Reproduction of *Sclerodermus sichuanensis*. *Insects* **2025**, *16*, 701. [CrossRef]
22. Wei, K.; Gao, S.; Tang, Y.; Wang, X.; Yang, Z. Determination of the Optimal Parasitoid-to-host Ratio for Efficient Mass-rearing of the Parasitoid, *Sclerodermus pupariae* (Hymenoptera: Bethylidae). *J. Appl. Entomol.* **2017**, *141*, 181–188. [CrossRef]
23. Yang, Z.; Wang, X.; Duan, Z.; Zhang, Y.; Zhang, Y.; Cao, L.; Wei, K. *Sclerodermus alternatus* (Hymenoptera: Bethylidae), a new Species from China, Parasitizing *Monochamus alternatus* (Coleoptera: Cerambycidae). *Zool. Syst.* **2024**, *49*, 258–266.
24. Yang, Z.; Wang, X.; Yao, Y.; Juli, R.; Cao, L. A new Species of *Sclerodermus* Latreille (Hymenoptera: Bethylidae) Parasitizing *Agrilus planipennis* (Coleoptera: Buprestidae) from China, with a Key to Chinese Species in the Genus. *Ann. Entomol. Soc. Am.* **2012**, *105*, 619–627. [CrossRef]
25. Jiang, X.; Zhou, Z. Research on the Female Reproductive System of *Sclerodermus sichuanensis* Xiao (Hymenoptera:Bethylidae). *J. Nanjing For. Univ.* **2006**, *30*, 95–97. (In Chinese)
26. Eliopoulos, P.; Stathas, G. Temperature-dependent Development of the Koinobiont Endoparasitoid *Venturia canescens* (Gravenhorst) (Hymenoptera: Ichneumonidae): Effect of Host Instar. *Environ. Entomol.* **2003**, *32*, 1049–1055. [CrossRef]
27. Li, Q.; Cheng, Y.; Luo, L.; Lei, C.; Jiang, X.; Zhang, L. Effect of Temperature on Parasitism Rate and Life History Parameters of a Solitary Endoparasitoid, *Zele chlorophthalmus* (Hymenoptera: Brachonidae). *Chin. J. Biol. Control.* **2017**, *33*, 575–583. (In Chinese)
28. Wang, X.; Wei, K.; Yang, Z.; Jennings, D.; Duan, J. Effects of Biotic and Abiotic Factors on Phenotypic Partitioning of Wing Morphology and Development in *Sclerodermus pupariae* (Hymenoptera: Bethylidae). *Sci. Rep.* **2016**, *6*, 26408. [CrossRef]

29. van Asch, M.; Tienderen, P.; Holleman, L.; Visser, M. Predicting Adaptation of Phenology in Response to Climate Change, an Insect Herbivore Example. *Glob. Change Biol.* **2007**, *13*, 1596–1604. [CrossRef]
30. Pelini, S.; Keppel, J.; Kelley, A.; Hellmann, J. Adaptation to Host Plants may Prevent Rapid Insect Responses to Climate Change. *Glob. Change Biol.* **2010**, *16*, 2923–2929. [CrossRef]
31. Felton, G.; Summers, C. Antioxidant Systems in Insects. *Arch. Insect Biochem.* **1995**, *29*, 187–197. [CrossRef]
32. Liu, X.; Fu, Z.; Kang, Z.; Li, H.; Liu, T.; Wang, D. Identification and Characterization of Antioxidant Enzyme Genes in Parasitiod *Aphelinus asychis* (Hymenoptera: Aphelinidae) and Expression Profiling Analysis under Temperature Stress. *Insects* **2022**, *13*, 447. [CrossRef] [PubMed]
33. Kang, Z.; Liu, F.; Xiang, L.; Yu, W.; Tan, X.; Zhang, S.; Tian, H.; Liu, T. The Potential Coordination of the Heatshock Proteins and Antioxidant Enzyme Genes of *Aphidius gifuensis* in Response to Thermal Stress. *Front. Physiol.* **2017**, *8*, 978. [CrossRef]
34. Feder, M.; Hofmann, G. Heat-shock Proteins, Molecular Chaperones, and the Stress Response: Evolutionary and Ecological Physiology. *Annu. Rev. Physiol.* **1999**, *61*, 243–282. [CrossRef]
35. Huang, L.; Chen, B.; Kang, L. Impact of Mild Temperature Hardening on Thermotolerance, Fecundity, and Hsp Gene Expression in *Liriomyza huidobrensis*. *J. Insect Physiol.* **2007**, *53*, 1199–1205. [CrossRef] [PubMed]
36. Le Bourg, É.; Valenti, P.; Lucchetta, P.; Payre, F. Effects of Mild Heat Shocks at Young age on Aging and Longevity in *Drosophila melanogaster*. *Biogerontology* **2001**, *2*, 155–164. [CrossRef] [PubMed]
37. Murphy, C.; McCrrroll, S.; Bargmann, C.; Fraser, A.; Kamath, R.; Ahringer, J.; Li, H.; Kenyon, C. Genes that Act Downsteam of DAF-16 to Influence the Lifespan of *Caenorhabditis elegans*. *Nature* **2003**, *424*, 277–283. [CrossRef]
38. Yi, J.; Wu, H.; Liu, J.; Lai, X.; Guo, J.; Li, D.; Zhang, G. Molecular Characterization and Expression of Six Heat Shock Pretein Genes in Relation to Development and Temperature in *Trichogramma chilonis*. *PLoS ONE* **2018**, *13*, e0203904. [CrossRef]
39. Tang, Y.; Wang, L.; Wang, Y.; Zhang, Y.; Wang, X.; Wei, K. Effects of Different Foundress Densities on Sex Ratio of the Offspring of Bethylid Wasps. *Sci. Silvae Sin.* **2022**, *58*, 161–168. (In Chinese)
40. Tang, Y.; Wang, L.; Wu, S.; Ou, B.; Wang, X.; Yang, Q.; Wei, K. Oviposition Pattern of *Sclerodermus alternatusi*: Benefits of Laying Unfertilized Eggs First. *Chin. J. Biol. Control.* **2020**, *36*, 832–836. (In Chinese)
41. Tang, Y.; Wang, L.; Zhang, Y.; Wu, S.; Wang, X.; Yang, Q. Effect of Different Parasitiod Colonies on the Mass Rearing of *Sclerodermus alternatus*(Hymenoptera: Bethylidae). *Sci. Silvae Sin.* **2020**, *56*, 97–103. (In Chinese)
42. Tang, Y.; Kang, K.; Chen, S.; Wang, L.; Che, H.; Wang, X.; Wei, K. Effects of High Temperature on the Progeny Development and *Wolbachia* Infection of *Sclerodermus sichuanensis* (Hymenoptera: Bethylidae). *Chin. J. Biol. Control.* **2023**, *39*, 752–759. (In Chinese)
43. Men, J.; Zhao, B.; Cao, D.; Wang, W.; Wei, J. Evaluating Host Location in Three Native *Sclerodermus* species and Their Ability to Cause Mortality in the Wood Borer *Aromia bungii* (Coleoptera: Cerambycidae) in Laboratory. *Biol. Control.* **2019**, *134*, 95–102. [CrossRef]
44. Huang, G.; Liu, Z.; Gu, S.; Zhang, B.; Sun, J. Identification and Functional Analysis of Odorant-binding Proteins of the Parasitoid wasp *Sclerodermus guani* Reveal a Chemosensory Synergistic Evolution with the Host *Monochamus alternatus*. *Int. J. Biol. Macromol.* **2023**, *249*, 126088. [CrossRef] [PubMed]
45. Abdi, M.; Jucker, C.; De Marchi, B.; Hardy, I.; Lupi, D. Performance of *Sclerodermus brevicornis*, a Parasitoid of Invasive Longhorn Beetles, when Reared on Rice Moth Larvae. *Entomol. Exp. Appl.* **2021**, *169*, 64–78. [CrossRef]

biology

MDPI

Article

Changes in the Microbiota of the Scale Insect (*Diaspis echinocacti*, Bouché, 1833) in *Opuntia stricta* Cladodes: Taxonomic and Metagenomic Analysis as a Function of Infestation Levels

Mikaelly Batista da Silva [1], Ana Beatriz Medeiros [2], Antonia Isabelly Monteiro dos Anjos [2], João Vitor Ferreira Cavalcante [3], Bianca Cristiane Ferreira Santiago [3], Shênia Santos Monteiro [1], Antonio Carlos Vital, Júnior [4], Rodrigo Juliani Siqueira Dalmolin [4], Hugo M. Lisboa [2,*] and Matheus Augusto de Bittencourt Pasquali [1,2,*]

1 Departamento de Engenharia e Gestão de Recursos Naturais, Universidade Federal de Campina Grande, Campina Grande 58401-490, Brazil
2 Departamento de Engenharia de Alimentos, Universidade Federal de Campina Grande, Campina Grande 58401-490, Brazil
3 Ambiente Multidisciplinar de Bioinformática—IMD, Universidade Federal do Rio Grande do Norte, Natal 59078-400, Brazil
4 Departamento de Bioquímica—CB, Universidade Federal do Rio Grande do Norte, Natal 59078-970, Brazil; rodrigo.dalmolin@imd.ufrn.br (R.J.S.D.)
* Correspondence: hugo.miguel@professor.ufcg.edu.br (H.M.L.); matheus.augusto@professor.ufcg.edu.br (M.A.d.B.P.)

Simple Summary: Insect pests can cause substantial damage to crops, and their associated microbiota may play a pivotal role in host physiology and plant interactions. In this study, we investigated the bacterial communities associated with the armored scale insect (*Diaspis echinocacti*) infesting *Opuntia stricta* in Brazil. By comparing insects collected from cladodes under low, medium, and high infestation levels, we assessed microbiota composition using *16S rRNA* gene-based analyses. Our results revealed a marked dominance of *Candidatus Uzinura diaspidicola* across all infestation levels, suggesting a critical symbiotic role in host survival. A potential decline in bacterial diversity with increasing infestation was observed, although this trend was not statistically robust given the limited sample size. These findings provide preliminary insights into the microbiota of scale insects and their possible involvement in host–plant interactions. Further studies employing larger sample sizes and integrative genomic approaches are required to validate these patterns and may ultimately inform microbiome-based strategies for sustainable cactus protection and food security.

Abstract: Drought-tolerant cactus *Opuntia stricta* sustains livestock in Brazil's semi-arid Northeast but suffers yield losses from the armored scale insect *Diaspis echinocacti*. Symbiotic bacteria are thought to underpin scale fitness; however, their response to pest pressure remains unexplored. We characterized the bacterial communities of *D. echinocacti* collected from cladodes displaying low, intermediate, and high infestation (n = 3 replicates per level) using 16S-rRNA amplicon sequencing, processed with nf-core/ampliseq. Shannon diversity declined from low to high density, and Bray–Curtis ordination suggested compositional shifts, although group differences were not significant (Kruskal–Wallis and PERMANOVA, $p > 0.05$). The obligate endosymbiont "*Candidatus Uzinura*" dominated all samples (>85% relative abundance) irrespective of density, indicating a resilient core microbiome. PICRUSt2 predicted a contraction of metabolic breadth at higher infestations, with convergence on energy- and amino acid biosynthesis pathways. Taken together,

 https://doi.org/10.3390/biology14091233

increasing pest density was associated with modest loss of diversity and functional streamlining, rather than wholesale turnover. These baseline data can guide future work on microbiome-based strategies to complement existing scale-insect control in dryland cactus systems.

Keywords: *Diaspis echinocacti*; 16S rRNA; endosymbionts; insect microbiota; *Candidatus Uzinura*

1. Introduction

Opuntia stricta, a cactus species native to Mexico, is highly adapted to semi-arid and arid regions. These regions cover approximately 37% of the Earth's surface and face significant challenges for agriculture and livestock because of high temperatures, irregular rainfall, and prolonged droughts. Known as fodder cactus, *O. stricta* serves as a vital feed resource for livestock during dry seasons, playing a fundamental role in agricultural sustainability, particularly in Brazil's semi-arid Northeast [1].

Beyond its nutritional value, the fodder cactus holds ecological, economic, and cultural significance by promoting food security, income generation, and sustainable production alternatives [2–4]. However, its cultivation is increasingly threatened by infestations of the armored scale insect *Diaspis echinocacti* (Hemiptera: Diaspididae), which attacks the cladodes and impairs plant growth and productivity. This pest is particularly harmful because of its high reproductive rate, rapid spread under arid climatic conditions, and the waxy carapace that affords females considerable resistance to insecticides [4,5].

Despite the agronomic importance of this pathosystem, no published study has yet examined how the microbiota of *D. echinocacti* changes along a gradient of infestation intensity on *O. stricta*, nor how such changes might modulate plant–insect interactions. This knowledge gap hampers the design of microbiome-informed control strategies that could complement classical biological or chemical measures.

The extensive and often indiscriminate use of chemical insecticides has already selected for resistant *D. echinocacti* populations and raised concerns about toxic residues in animal products derived from livestock fed on infested cacti [6]. Understanding the mechanisms underlying pest adaptation and survival, particularly the role of microbial symbionts, is therefore crucial. Similar to many insects, scale insects harbor both endo- and ectosymbiotic microorganisms that provide essential nutrients, modulate immunity, and influence host development [7–9].

Armored scale insects, including *D. echinocacti*, rely on obligate bacterial endosymbionts housed within specialized bacteriocytes that form the bacteriome, a dedicated organ for hosting intracellular symbionts. The primary endosymbiont, *Candidatus Uzinura diaspidicola*, provides essential amino acids and other nutrients critical for host survival on nutrient-poor plant sap, rather than residing in the gut or shifting with feeding behavior [10]. This symbiotic relationship, first characterized by Gruwell et al. [11], is a hallmark of Diaspididae, with *Uzinura* showing high genomic integration with its host, including gene loss and compensatory metabolic pathways. Sabree et al. [12] further confirmed *Uzinura's* role in nutrient provisioning, particularly in metabolizing plant-derived compounds. Additional studies, such as Szklarzewicz et al. [13] and Bosch et al. [14], highlight the structural and functional stability of bacteriome-hosted endosymbionts in scale insects, though secondary symbionts may modulate immunity or detoxify plant defenses. Understanding these microbial associations is critical for elucidating pest fitness and developing targeted control strategies.

High-throughput sequencing of the *16S rRNA* gene has revolutionized the study of insect-associated microbiota by providing an efficient and cost-effective snapshot of community structure [15]. Nonetheless, analytical choices strongly influence downstream ecological inferences; benchmark comparisons—such as the multi-pipeline assessment by Straub et al. [16]—have shown that different amplicon workflows can bias estimates of diversity and taxon prevalence. In the present work, we therefore applied the nf-core/ampliseq pipeline, which performed favorably in that benchmark, while remaining aware of its limitations.

Although interest in pest-associated microbiota is growing, most recent studies have focused on other Hemipterans (e.g., *Puto barberi*; [17]) or the efficacy of control agents such as detergents and biopesticides against mealybugs (*Drosicha mangiferae*; [18]) without integrating the plant–insect–microbiota triad. Consequently, how infestation intensity shapes *D. echinocacti* symbioses—and how this, in turn, might influence *O. stricta* defense responses—remains poorly characterized.

We therefore hypothesized that microbiota diversity would decline and functional predictions would converge as the intensity of *D. echinocacti* infestation increases on *O. stricta*. Testing this hypothesis will provide baseline information for developing sustainable, microbiome-aware management strategies that reduce chemical dependency and support fodder cactus production in semi-arid regions.

2. Materials and Methods

2.1. Sample Collection

The insects used in this study were collected from *Opuntia stricta* cladodes from the experimental area of the Instituto Nacional do Semiárido (INSA), located in Campina Grande, Paraíba, Brazil (GPS coordinates: latitude: −7.252411, longitude: −35.946111), in June 2024. The cladodes were classified into three categories based on visual estimation of armored scale insect: Group 1 (low infestation, <20% surface coverage), Group 2 (intermediate infestation, 20–50% surface coverage), and Group 3 (high infestation, >50% surface coverage) (Figure 1). Three biological replicates were collected per infestation level, each consisting of approximately 50 insects pooled from a single cladode to ensure sufficient DNA yield and account for individual variability. Cladodes were selected from different *O. stricta* plants, spaced at least 5 m apart, to ensure spatial independence. No temporal replication was performed due to logistical constraints. Although the selected plants were of similar age, the cladodes varied in age. Cladode age, insect age, and environmental variables such as microhabitat (e.g., sun exposure, soil conditions) and sampling time were not quantitatively controlled in this study. Older cladodes typically exhibit higher levels of *D. echinocacti* infestation [19]. These factors may represent confounding variables that influence the composition of the microbial community. Due to logistical constraints, quantitative assessments and controlled sampling designs were not implemented.

The insects were carefully removed from the *O. stricta* cladodes using forceps, placed in sterile 1.5 mL microcentrifuge tubes, and immediately stored at −20 °C to preserve microbial DNA during transport. The samples were transferred to the laboratory within 24 h, where DNA extraction was performed. Storage at −20 °C was chosen due to logistical constraints that prevented the use of −80 °C facilities. Although immediate extraction or preservation at −80 °C is generally recommended for low-biomass insect samples to reduce potential DNA degradation and changes in microbial community composition, the conditions employed were sufficient to maintain DNA integrity for subsequent analyses.

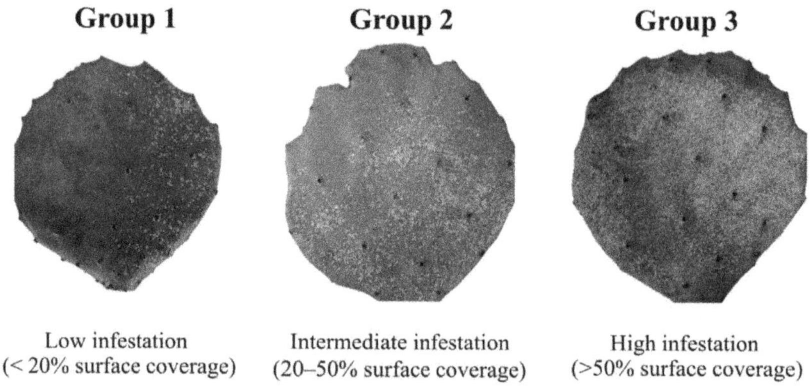

Figure 1. Cladode of *O. stricta* infested by *Diaspis echinocacti*, showing three infestation levels: Group 1 (low infestation, <20% surface coverage), Group 2 (intermediate infestation, 20–50% surface coverage), and Group 3 (high infestation, >50% surface coverage). The magnified detail reveals numerous scale insects adhering to the cladode surface, exhibiting the characteristic circular morphology of the species' shields. Source: Photo taken by the author.

Surface decontamination of *D. echinocacti* was performed by gently brushing the insects with a sterile soft-bristled brush under a stereomicroscope to remove superficial field debris, such as dust or plant residues, without damaging their delicate structure (approximately 1–2 mm in diameter) or waxy shields. Chemical sterilization methods, such as rinsing with 70% ethanol or 1% sodium hypochlorite, were tested but avoided, as they caused physical damage (e.g., dislodgement of the waxy shield or cuticle degradation) or dragged insects due to surface wetting, compromising sample integrity. For example, ethanol rinsing led to partial shield detachment, potentially exposing internal tissues, while sodium hypochlorite caused visible cuticle erosion. The soft brush method minimized damage while reducing external contaminants, though we acknowledge that it may not eliminate all plant-associated or environmental bacteria, which could influence microbiome profiles. Future studies should explore optimized, non-damaging decontamination protocols for small armored scale insects to further reduce external bacterial interference.

2.2. *Procedures for 16S Sequencing on the iSeq 100*

2.2.1. DNA Extraction and 16S Amplification

DNA was extracted using 3 mg of the insect using the phenol–chloroform method (pH 8.0). Samples were manually macerated with a sterile pestle directly in the microtubes until complete homogenization in the lysis solution, without the use of bead-beating or other mechanical disruption methods. Extraction blanks (negative controls without biological material) were included in all procedures, and no contamination was detected.

To amplify the bacterial 16S, 5 µL of 5X Colorless GoTaq® Flexi Buffer, 0.2 µL of GoTaq® G2 Hot Start Polymerase (Promega, Madison, WI, USA), 1 µL of dNTPs mix (5 mM) (Promega, Madison, WI, USA), 1.6 µL of $MgCl_2$, 2 µL of Forward + Reverse primer (5 µM) (synthesized by Integrated DNA Technologies—IDT, Coralville, IA, USA) and 1 µL of extracted DNA were placed in each 0.2 mL microtube for a total volume of 25 µL. The tubes were then placed in the ABI9700 thermal cycler (Applied Biosystems, Foster City, CA, USA) with the following program: 94 °C for 3 min, followed by 40 cycles of 94 °C for 1 min, 53 °C for 1 min, and 72 °C for 2 min. After the 40 cycles, there was a step of 72 °C for 5 min. To check for amplification, the amplicons were subjected to agarose gel electrophoresis (1.5%).

The primer sequences used were as follows:

16S Amplicon PCR Forward Primer = 5′-3′
TCGTCGGCAGCGTCAGATGTGTATAAGAGACAGCCTACGGGNGGCWGCAG
16S Amplicon PCR Reverse Primer = 5′-3′
GTCTCGTGGGCTCGGAGATGTGTATAAGAGACAGGACTACHVGGGTATCTAATCC

2.2.2. Purification and PCR Binding INDEX

The 16S amplicons were purified using AMPure XP magnetic beads (Beckman Coulter, Brea, CA, USA) according to the manufacturer's recommendations. PCR was then carried out to bind the indexes using the Nextera XT Index kit (Illumina, San Diego, CA, USA). For this, 5 µL of each purified amplicon, 25 µL of 2X PCR BIO Ultra Mix (PCR Biosystems, Wayne, PA, USA), 5 µL of the respective XT index 1 primer (N7xx), and 5 µL of the respective XT index 2 primer (S5xx) were placed in each 0.2 mL microtube for a final volume of 40 µL. The tubes were then placed in the ABI9700 thermal cycler (Applied Biosystems, Foster City, CA, USA) with the following program: 15 °C for 1 min, 95 °C for 3 min, followed by 8 cycles of 95 °C for 30 s, 55 °C for 30 s, and 72 °C for 30 s. After the 8 cycles, there was a step of 72 °C for 5 min. To check for index binding, the amplicons were subjected to agarose gel electrophoresis (1.5%). A second purification was performed after index binding using AMPure XP magnetic beads (Beckman Coulter, Brea, CA, USA) according to the manufacturer's instructions. Finally, the samples were quantified using the Quantus fluorometer (Promega, Madison, WI, USA) and normalized before being placed on the cartridge for sequencing.

2.2.3. Bioinformatics Analysis

The data, with an average of 118,666 raw reads per replicate, averaging 326,547 for each group, were processed using version 2.10.0 of the nf-core/ampliseq pipeline [16]. For the inference of amplicon sequence variants (ASV), DADA2 [20] was employed, while QIIME2 [21] was used to calculate diversity indices. Following the pipeline's workflow, reads were first dynamically truncated based on quality (at the first base where the median quality score dropped below 25, while retaining at least 75% of reads) and filtered to discard sequences with a length shorter than 50 bp. Chimeric sequences were then removed using DADA2's "consensus" method. Taxonomic assignments for the resulting ASVs were made against the SILVA version 138 reference database. Additionally, any ASVs classified as "mitochondria" or "chloroplast" were excluded from the dataset before diversity indices were calculated using QIIME2. This quality control process retained an average of 91.7% of the initial reads.

Taxonomic count tables were obtained by summing the read counts of all ASVs assigned to a given taxon. To assess statistical differences in community composition (beta diversity), a permutational multivariate analysis of variance (PERMANOVA) was performed with 999 permutations. For alpha diversity, group significance was determined using a Kruskal–Wallis test with 2 degrees of freedom, and the resulting pairwise comparisons were adjusted for multiple testing using the False Discovery Rate (FDR) correction. Functional predictions were generated with PICRUSt2 [22] based on *16S rRNA* amplicon sequence variants (ASVs), inferring possible metabolic pathways, such as amino acid biosynthesis and energy metabolism. These predictions represent rough approximations, as accuracy depends on the availability of closely related reference genomes. In communities dominated by poorly characterized endosymbionts, such as *Canidatus Uzinura diaspidicol*, functional inference may be interpreted as hypotheses that require validation through metagenomic or biochemical approaches. All original code has been deposited on GitHub and is publicly available at https://github.com/dalmolingroup/cochineal_16s.

3. Results

3.1. Analysis of Biological Diversity

The Shannon index indicated a potential trend of higher microbial diversity in the low infestation group (Group 1), with a gradual decrease observed in Groups 2 and 3 (Figure 2). Samples were analyzed in triplicate, accounting for possible developmental variations among individuals. However, the Kruskal–Wallis test showed no statistically significant differences among groups ($p > 0.05$), likely due to the small sample size ($n = 3$ per group), which limits statistical power. These observations are hypothesis-generating and suggest that microbial diversity may decrease with increasing infestation, warranting further investigation with larger sample sizes to confirm biological significance.

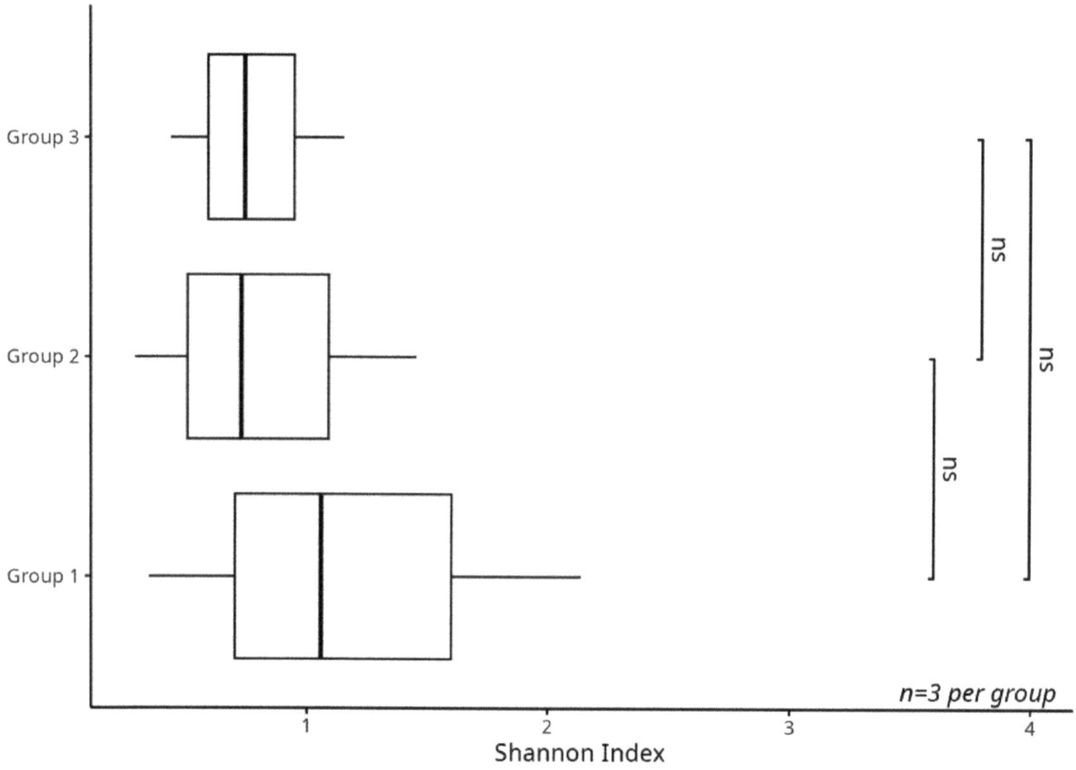

Figure 2. Alpha diversity of the microbiota associated with *D. echinocacti* under different infestation levels, based on the Shannon index. A higher diversity is observed in Group 1 (low infestation), followed by Group 2 (intermediate), and the lowest in Group 3 (high infestation). These results indicate a decline in microbial diversity with increasing infestation, suggesting a potential process of functional specialization. ns: not significant.

In the early stages of infestation, the insect may rely on a broader microbial repertoire to counter host plant defenses. As the infestation progresses, the microbiota becomes increasingly specialized, thereby enhancing the insect's ability to adapt and become resilient. These microbial shifts likely reflect functional changes aligned with host needs, as reported by Chen et al. [23]. Similarly, Yun et al. [24] demonstrated that the composition of insect gut microbiota is shaped by host development, diet, environment, and phylogeny—factors that not only modulate the presence but also influence microbial function throughout the insect's life cycle.

Samples were analyzed in triplicate, accounting for possible developmental variations among individuals. Although trends differed between groups, the Kruskal–Wallis test revealed no significant statistical differences ($p > 0.05$), underscoring the need for further integrative studies on insect behavior and microbiota dynamics during infestation.

3.2. Taxonomic Composition of the Microbiota Associated with Scale Insects

The taxonomic composition of the microbiota associated with scale insects was assessed based on both the relative and absolute abundance of taxa, allowing for a comprehensive analysis of differences between infestation groups. The study of relative abundance revealed that *Candidatus Uzinura* was the dominant taxon across all groups, consistent with previous reports identifying it as a common endosymbiont in armored scale insects [12]. The designation "Candidatus" refers to bacteria that, although genetically determined, cannot yet be cultured using conventional laboratory methods. In this study, *C. Uzinura* accounted for 86.4% of the microbial community in Group 1 (low infestation), 94.2% in Group 2 (intermediate infestation), and 92.4% in Group 3 (high infestation), with no statistically significant differences among the groups (Figure 3).

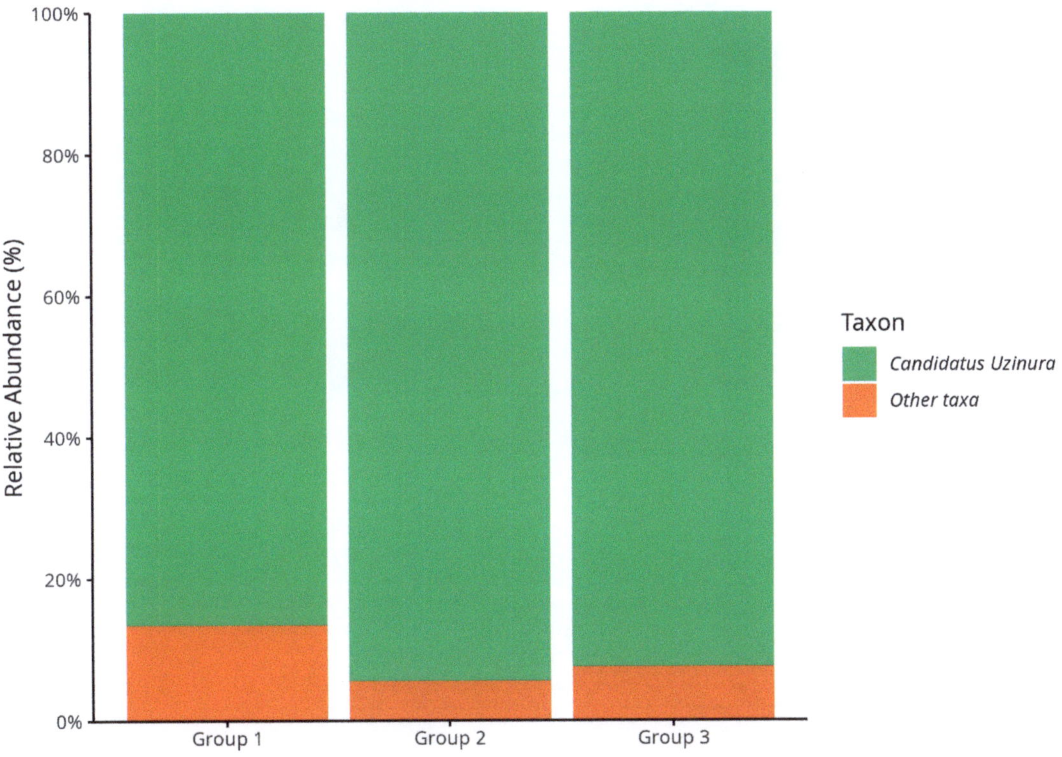

Figure 3. Relative abundance of *C. Uzinura* and other bacterial taxa in the microbiota of *D. echinocacti* across different infestation levels. *C. Uzinura* was the dominant taxon in all groups, especially in Groups 2 and 3 (intermediate and high infestation), where it accounted for nearly the entire microbial community. Group 1 (low infestation) displayed a higher proportion of other taxa, indicating greater microbial diversity at earlier stages of infestation.

The consistent dominance of *Candidatus Uzinura* across all groups confirms its established role as an obligate endosymbiont in Diaspididae, providing essential nutrients such as amino acids critical for the host's survival on nutrient-poor plant sap [12]. This aligns with findings by Gruwell et al. [11], who identified *Uzinura* as the primary nutritional symbiont in armored scale insects, and Sabree et al. [12], who demonstrated its metabolic integration with the host. Moreover, the proportion of "other taxa" decreased from Group 1 to Group 2, with a slight increase in Group 3. Although subtle, this trend supports the hypothesis that the microbiome becomes less diverse and increasingly dominated by a single taxon as infestation progresses, characterizing a process of symbiotic specialization.

To specifically investigate the composition of the less abundant community members, an analysis was performed on the dataset after computationally removing all reads from the overwhelmingly dominant endosymbiont, *Candidatus Uzinura*. The results of this targeted analysis (Figure 4) show that the phylum Proteobacteria was the most abundant in all groups, particularly in Group 1, followed by Groups 3 and 2. Group 1 showed the highest overall microbial abundance, with notable representation of other phyla such as Actinobacteriota, Acidobacteriota, and Bacteroidota, suggesting a more complex and diverse microbiota at the early stages of infestation. In contrast, Group 2 exhibited a significant reduction in both total abundance and phylum diversity. Group 3 showed partial recovery of total abundance but was dominated by fewer phyla, indicating a more specialized microbial community.

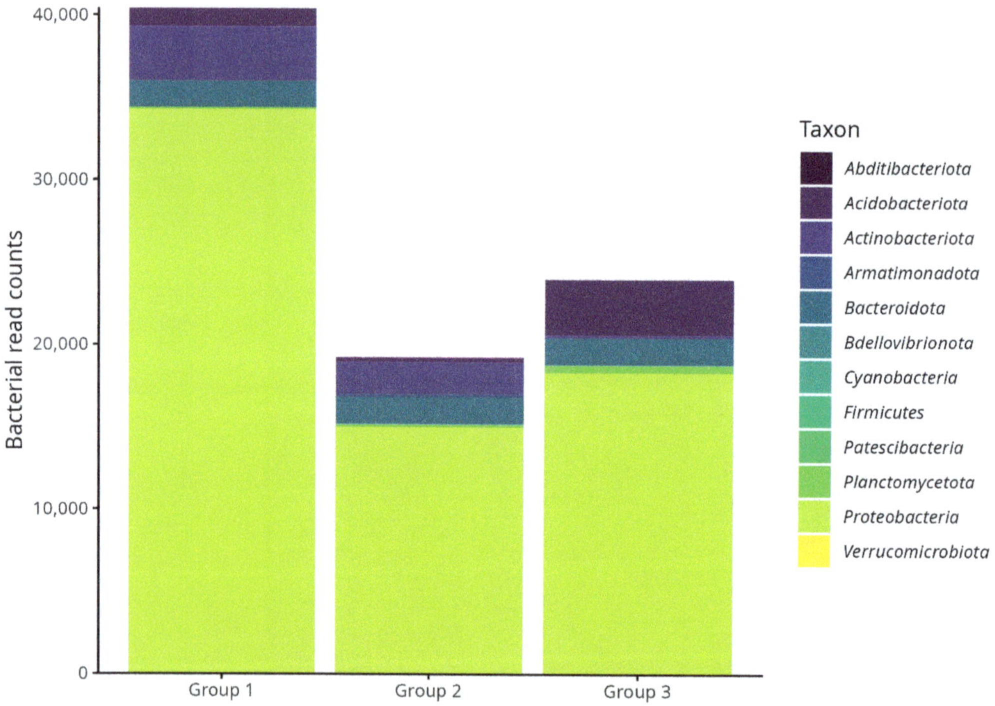

Figure 4. Bacterial read counts at the phylum level associated with the armored scale insect *D. echinocacti* after computational removal of the dominant endosymbiont *Candidatus Uzinura diaspidicola*, shown across different infestation levels (Group 1: low; Group 2: intermediate; Group 3: high). Within this non-symbiont portion of the community, the phylum Proteobacteria was predominant in all groups, especially in Group 1, followed by Acidobacteriota and Bacteroidota. A general decrease in total bacterial read counts was observed from Group 1 to Group 2, with a slight increase in Group 3.

These findings support the hypothesis that microbial diversity and composition are modulated by infestation level. A richer and taxonomically varied microbiota at early infestation stages may aid in insect adaptation and interaction with the host plant. As infestation intensifies, the microbial community becomes simplified, favoring specific symbionts better suited to the host's stress conditions.

3.3. Functional Profile of the Microbiota Associated with Scale Insects

The functional potential of the microbiota associated with *D. echinocacti* was inferred using the PICRUSt2 tool, which applies various algorithms to predict gene function based on *16S rRNA* gene sequencing data. The 30 most abundant metabolic pathways across all samples, as expected and referenced in the MetaCyc database [25], are shown in Figure 5.

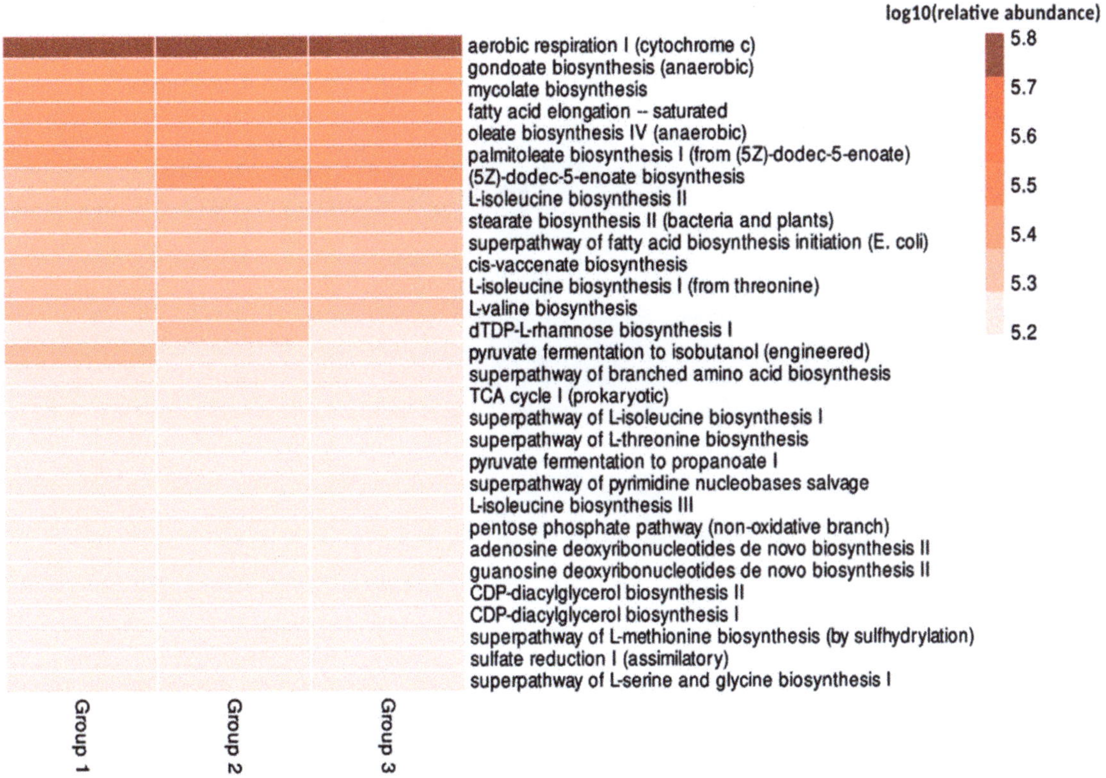

Figure 5. Heatmap showing the 30 most abundant metabolic pathways inferred from the microbiota associated with *D. echinocacti* across three groups of infestation intensity in *O. stricta*. The color gradient represents the $\log_{10}$-transformed relative abundance of each metabolic pathway (ranging from 5.2 to 5.8, where darker shades indicate higher abundance). Pathways such as (5Z)-dodec-5-enoate biosynthesis and dTDP-L-rhamnose biosynthesis I show higher abundance in Groups 1 and 2, respectively, reflecting potential functional specialization. Group 1: low infestation; Group 2: moderate infestation; Group 3: high infestation.

PICRUSt2-inferred functional profiles suggest a predominance of metabolic pathways related to energy production and biosynthesis, potentially consistent with the genomic capabilities of *Candidatus Uzinura diaspidicola* described by Sabree et al. [12]. Pathways such as aerobic respiration I (cytochrome c), anaerobic gondoate biosynthesis, mycolate biosynthesis, and fatty acid elongation processes showed the highest $\log_{10}$-transformed inferred read counts (ranging from 5.2 to 5.8, where higher values indicate greater relative abundance) across all groups, particularly in Group 1 (low infestation). Aerobic respiration I was inferred in all groups, suggesting a conserved metabolic role. In Group 1, the (5Z)-dodec-5-enoate biosynthesis pathway had higher predicted abundance (log_{10} value ~ 5.8), hypothesized to support early-stage colonization through potential microbial signaling or immune enhancement [26,27]. Group 2 exhibited elevated predicted abundance of the dTDP-L-rhamnose biosynthesis I pathway (log_{10} value ~ 5.7), hypothesized to be linked to polysaccharide production for bacterial colonization during intermediate infestation. Group 3 exhibited no significantly elevated predicted pathways, suggesting functional stabilization. These inferences, based on 16S rRNA data, are hypothetical due to the limitations of PICRUSt2, particularly for endosymbionts like *Uzinura* with potentially limited reference genomes, and require validation with metagenomic or biochemical studies.

The predicted functional potential of the microbiota, inferred using PICRUSt2, aligns with the genomic capabilities of *Candidatus Uzinura diaspidicola* described by Sabree et al. [12], which include pathways for synthesizing essential amino acids (e.g., L-isoleucine, L-valine) and fatty acids critical for host nutrition on nutrient-poor plant sap. The predominance of biosynthetic pathways associated with amino acids such as L-isoleucine, L-valine, and L-methionine, especially in Group 1, reflects the metabolic versatility of the microbiota during early infestation stages. This group exhibited a more diverse and abundant functional repertoire compared to Groups 2 and 3, indicating greater metabolic versatility at early stages of infestation. In contrast, the functional diversity in Groups 2 and 3 appeared reduced, reflecting a more specialized or simplified microbial community.

These findings are consistent with the taxonomic results, which revealed greater microbial diversity in Group 1 and dominance of specific taxa in Groups 2 and 3. The predicted reduction in metabolic diversity in samples with higher infestation levels reinforces the hypothesis of a functionally specialized microbiota, possibly co-evolved with the host under more restrictive ecological conditions.

4. Discussion

The analysis of alpha diversity in the microbiota associated with *D. echinocacti* revealed a decreasing trend across infestation levels in *O. stricta*. The highest Shannon index was observed in the low-infestation group (Group 1), followed by a gradual decline in Groups 2 and 3. This trend suggests that the microbial community, including the bacteriome-hosted *Candidatus Uzinura* and secondary symbionts, undergoes functional modulation in response to the host insect's physiological demands, consistent with an adaptive strategy. At early infestation stages, microbial diversity appears to play a crucial adaptive role, supporting the insect in overcoming the plant's structural and chemical defenses. Symbionts involved in detoxifying secondary metabolites, modulating immunity, and aiding nutrient absorption are essential [24,28]. However, these trends were not statistically significant ($p > 0.05$), likely due to the small sample size ($n = 3$ per group), which limits statistical power.

However, it is important to note that cladode age, which correlates with infestation intensity, may also influence microbial composition, as alder cladodes tend to host more established insect colonies. This confounding factor was not controlled in the present study, and future work should incorporate cladode or insect age as a variable to better clarify its effects. In addition, sample storage at −20 °C for 24 h prior to DNA extraction may have introduced minor alterations in microbial community profiles. Future studies should preferentially employ immediate extraction or −80 °C storage to ensure optimal preservation. Finally, as chemical sterilization methods could damage insect integrity, surface decontamination was performed using a soft brush. While this approach minimizes structural damage, it may have allowed residual plant- or environment-associated bacteria to remain;

thus, the development of non-destructive sterilization protocols should be considered in future investigations. Furthermore, functional predictions from PICRUSt2, such as roles in biosynthesis or plant defense neutralization, are coarse approximations, particularly unreliable for endosymbionts like *Uzinura* due to limited reference genomes, and require validation with metagenomics, metatranscriptomics, or targeted assays (e.g., qPCR of functional genes). Future studies should include controls like extraction blanks and plant microbiome analyses to account for external bacteria and employ statistical methods like ANCOM or qPCR-based quantitation to improve compositional and abundance analyses.

This diverse community typically includes generalist bacteria capable of metabolizing plant exudates and neutralizing broad-spectrum defense compounds [29]. As infestation intensifies, a shift toward microbiota specialization is observed, with dominance of symbionts optimized for nutrient extraction and host homeostasis maintenance [6]. This transition aligns with the idea that symbiotic composition adapts to the specific nutritional and defensive context of the host [30]. The stable dominance of *Candidatus Uzinura* in the bacteriome across infestation levels underscores its role as an obligate endosymbiont in Diaspididae, providing essential nutrients and potentially aiding in detoxification of plant defenses, as observed in other scale insects [12,13]. Unlike gut-associated microbiota, which may shift with feeding behavior, *Uzinura's* intracellular presence in the bacteriome ensures consistent metabolic support regardless of host feeding dynamics. In our study, the reduction in microbial diversity at advanced infestation stages may indicate ecological specialization, which favors functions such as sustained nutrient uptake, immune modulation, and resistance to cumulative defensive compounds, including alkaloids, phenolics, and saponins, produced by cacti [29,31]. These shifts suggest functional adaptations linked to insect development and infestation intensity. Although no statistically significant differences were found (Kruskal–Wallis, $p > 0.05$), the observed trends suggest biologically relevant patterns that may be obscured by individual variability or limitations in sample size. Triplicate analyses were employed to mitigate intraspecific variation; however, future studies with increased statistical power and integrated metagenomics may provide further clarification of these mechanisms.

Microbiota analysis, based on both relative and absolute abundances, revealed changes that aligned with infestation levels, suggesting shifts in ecological and functional community structure. *C. Uzinura* dominated all groups (Figure 3), accounting for 86.4% in Group 1, 94.2% in Group 2, and 92.4% in Group 3. This consistency confirms its role as an obligate symbiont in Diaspididae scale insects [12], likely contributing essential nutrients absent from plant sap [32]. Though primarily nutritional, *Candidatus Uzinura diaspidicola* is hypothesized to potentially contribute to neutralizing plant secondary metabolites in *O. stricta*, based on PICRUSt2-inferred pathways [13,33]. Given PICRUSt2′s limitations for endosymbionts, this role is speculative and requires validation with metagenomics or qPCR-based assays. Its metabolic integration with the insect host is supported by complementary pathways and mutual gene loss and compensation [34,35]. As infestation progresses, changes in plant chemical defenses may require microbiota co-adaptation [36]. A higher proportion of "other taxa" in Group 1, decreasing in later groups, suggests a trend toward specialization. As infestation advances, microbial diversity decreases, favoring *Candidatus Uzinura diaspidicola* in the bacteriome and select secondary symbionts well-adapted to the internal host environment [12,13]. This may reflect ecological filtering, where more efficient mutualist outcompetes less beneficial taxa. The diversity of insect feeding habits and habitats makes it challenging to define a "core" microbiome beyond obligate endosymbionts like *Uzinura*, though specific phyla, including Proteobacteria, Bacteroidetes, Firmicutes, Actinomycota, Spirochaetota, and Verrucomicrobiota, are consistently found across insect taxa [37,38]. Absolute abundance analysis (Figure 4) confirmed that Proteobacteria were the dominant group in all samples, with Group 1 exhibiting the highest representation and most remarkable phylogenetic diversity. This rich, early-stage community likely supports colonization by facilitating key metabolic processes [22]. Other phyla, including Actinobacteriota, Acidobacteriota, and Bacteroidota, were also more prominent in Group 1.

Proteobacteria are involved in degrading organic matter, fermenting sugars, and synthesizing vitamins and aromatic compounds, enhancing nutrient assimilation [39]. Their dominance in various insects, including *Bombyx mori* and *Spodoptera litura*, underscores their central role in the insect microbiome [40–42]. Bacteroidetes specialize in breaking down complex polysaccharides, such as cellulose, which facilitates the digestion of lignocellulosic material [43,44]. In *Cyrtotrachelus buqueti*, for example, Bacteroidota contribute glycoside hydrolases that release simple sugars from bamboo-derived polysaccharides. Their presence in *D. echinocacti* suggests a role in digesting complex plant compounds.

Actinobacteria, known for antimicrobial activity, were more abundant in Group 1. These bacteria may support digestion and defense during active feeding stages, as shown in moth larvae [45]. In contrast, Group 2 exhibited a drop in microbial abundance and diversity, likely reflecting physiological stress and disruption of mutualistic associations [22]. Group 3 exhibited partial recovery, but with a dominance of a few phyla, suggesting symbiotic specialization under stress [46]. Overall, the microbiota composition and structure in *D. echinocacti* appear to be modulated by infestation intensity. A diverse early-stage microbiome may promote colonization, while advanced stages involve functional streamlining centered on key symbionts. Endosymbiotic bacteria demonstrate efficient energy metabolism, supported by modular respiratory chains that enable adaptation to host-associated environments [47]. Predictive functional analysis identified the "aerobic respiration I (cytochrome c)" pathway across all groups, suggesting a conserved metabolic core crucial for energy generation. The microbiota also plays roles in immunity, reproduction, and stress resistance [22]. Group-specific pathways further suggest adaptation to host physiology. In Group 1, the exclusive detection of (5Z)-dodec-5-enoate biosynthesis—an unsaturated medium-chain fatty acid—may indicate microbial signaling or immune-enhancing functions [26,27]. In Group 2, the detection of the "dTDP-L-rhamnose I biosynthesis" pathway suggests the production of structural polysaccharides essential for bacterial colonization and immune modulation.

Common detection of amino acid biosynthesis (e.g., L-isoleucine, L-valine, L-threonine), fatty acid elongation, and oleate biosynthesis across groups reflects metabolic functions geared toward cellular homeostasis and membrane stability [48]. In contrast, no exclusive pathways were found in Group 3, suggesting functional consolidation around essential symbiotic functions and reduced redundancy. These findings support the hypothesis of functional succession in the microbiota of *D. echinocacti*, which is shaped by infestation level and the physiological demands of the host insect. Similar functional stability of bacteriome-hosted endosymbionts has been observed in other scale insects, where *Candidatus Uzinura* and related symbionts maintain consistent metabolic roles across developmental stages to support host nutrition and adaptation to plant defenses [12,13,35]. These patterns highlight the conserved role of obligate endosymbionts in Diaspididae, distinct from the more dynamic gut microbiota of other insects. Ultimately, interactions between insects and their endosymbionts drive coevolution, with high genomic integration. In some phloem-feeding insects, key metabolic pathways are partitioned between host and symbiont genomes—a phenomenon known as "collaborative pathways"—which are still largely unexplored in insects with variable diets [49].

5. Conclusions

The microbiota of *D. echinocacti* in *O. stricta* shows a potential trend of decreasing diversity with increasing infestation intensity, suggesting possible functional specialization, though these trends were not statistically significant ($p > 0.05$) due to the small sample size ($n = 3$ per group). *Candidatus Uzinura diaspidicola* dominates across all groups, potentially providing essential nutrients and hypothesized to aid in neutralizing plant defenses. These hypothesis-generating findings highlight the potential for microbiome-based pest management strategies in cactus cultivation. Future studies should use larger sample sizes and control for confounders like cladode/insect age and environmental factors (e.g., microhabitat, sampling time) through covariate modeling, stratified sampling by cladode age, or controlled experiments to confirm these trends and elucidate mechanisms of insect–microbiota–plant interactions.

Author Contributions: Conceptualization, S.S.M. and M.A.d.B.P.; data curation, S.S.M.; formal analysis, A.I.M.d.A. and S.S.M.; funding acquisition, H.M.L. and M.A.d.B.P.; investigation, M.B.d.S., A.B.M., J.V.F.C. and B.C.F.S.; methodology, R.J.S.D. and M.A.d.B.P.; resources, R.J.S.D. and H.M.L.; supervision, R.J.S.D. and M.A.d.B.P.; validation, A.C.V.J.; writing—original draft, M.B.d.S. and H.M.L.; writing—review and editing, H.M.L. and M.A.d.B.P. All authors have read and agreed to the published version of the manuscript.

Funding: FAPESQ-PB EDITAL UNIVERSAL 2021-09; CNPq 306165/2023-6.

Institutional Review Board Statement: The animal study protocol was approved by the Institutional Ethics Committee of Universidade Federal Campina Grande protocol code CAAE: 15354119.3.0000.5182 and with date of approval 08/2020.

Informed Consent Statement: Not applicable.

Data Availability Statement: The raw sequencing data generated in this study have been deposited in the NCBI Sequence Read Archive (SRA) under BioProject accession number PRJNA1309975 (https://www.ncbi.nlm.nih.gov/bioproject/PRJNA1309975).

Acknowledgments: The authors thank the technical team of the National Institute of the Semi-Arid Region (INSA) for their support and availability in providing the samples studied and the team at the Genaptus Laboratory for their assistance with sequencing analyses. We also express our gratitude to the Coordination for the Improvement of Higher Education Personnel (CAPES), as well as the funding agencies that supported this research, and to the Graduate Program in Engineering and Management of Natural Resources—UFCG. The authors would also like to thank NPAD/UFRN for computational resources. A special thanks to colleagues and reviewers whose contributions were essential for improving this manuscript.

Conflicts of Interest: The authors declare no conflicts of interest.

References

1. Albuquerque Junior, P.S.; Silva, C.A.D.; Ramos, R.S.; Zanuncio, J.C.; Castellani, M.A. *Diaspis echinocacti* (Hemiptera: Diaspididae) on Cactus Pear Cladodes: Biological Aspects at Different Temperatures. *Braz. J. Biol.* **2023**, *83*, e274016. [CrossRef] [PubMed]
2. Dubeux, J.C.B.; Santos, M.V.F.; Souza, R.T.A.; Siebert, A. Cactus: The New Green Revolution in Drylands. *Acta Hortic.* **2022**, 233–240. [CrossRef]
3. Marengo, J.A.; Galdos, M.V.; Challinor, A.; Cunha, A.P.; Marin, F.R.; Vianna, M.d.S.; Alvala, R.C.S.; Alves, L.M.; Moraes, O.L.; Bender, F. Drought in Northeast Brazil: A Review of Agricultural and Policy Adaptation Options for Food Security. *Clim. Resil. Sustain.* **2022**, *1*, e17. [CrossRef]
4. Rodrigues, P.G.; Garcez, D.S.M.; Silva, C.M.; Santana, C.C.S.; Santana, J.C.S.; Lopes, C.D.C.; Muniz, E.N.; Oliveira Júnior, G.M.; de Moura, R.S.; Souza, J.C. Use of Palm Bran (*Nopalea cochenillifera* (L.) Salm-Dyck) in Partial Replacement of Concentrate in Maintenance Equine Diets—A Pilot Study. *Arch. Anim. Breed.* **2021**, *64*, 273–282. [CrossRef]

5. Dantas, P.C.; de Araújo, R.G.V.; de Abreu, L.A.; Sabino, A.R.; dos Santos Silva, C.; Figueiroa, L.E.; Cunha, J.L.X.L.; Duarte, A.G. Avaliação de Extratos Botânicos No Controle Da Cochonilha de Escama *Diaspis Echinocacti* (Brouché, 1833) (Hemiptera: Diaspididae). *Braz. J. Dev.* **2019**, *5*, 2012–2017.
6. El Aalaoui, M.; Rammali, S.; Bencharki, B.; Sbaghi, M. Integrated Biological Control of *Diaspis Echinocacti* (Bouché) on Opuntia Ficus-Indica (L.) Mill (Cactaceae) Using Predatory Ladybirds and Fungal Pathogens. *Crop Prot.* **2025**, *187*, 106950. [CrossRef]
7. Bustamante-Brito, R.; Vera-Ponce de León, A.; Rosenblueth, M.; Martínez-Romero, E. Comparative Genomics of the Carmine Cochineal Symbiont Candidatus Dactylopiibacterium Carminicum Reveals Possible Protection to the Host against Viruses via CRISPR/Cas. *Syst. Appl. Microbiol.* **2024**, *47*, 126540. [CrossRef]
8. Zepeda-Paulo, F.; Romero, V.; Celis-Diez, J.L.; Lavandero, B. A Newly Discovered Bacterial Symbiont in the Aphid Microbiome Identified through 16S RRNA Sequencing. *Symbiosis* **2024**, *93*, 223–228. [CrossRef]
9. Lange, C.; Boyer, S.; Bezemer, T.M.; Lefort, M.-C.; Dhami, M.K.; Biggs, E.; Groenteman, R.; Fowler, S.V.; Paynter, Q.; Verdecia Mogena, A.M.; et al. Impact of Intraspecific Variation in Insect Microbiomes on Host Phenotype and Evolution. *ISME J.* **2023**, *17*, 1798–1807. [CrossRef] [PubMed]
10. Gruwell, M.E.; Flarhety, M.; Dittmar, K. Distribution of the Primary Endosymbiont (Candidatus Uzinura Diaspidicola) Within Host Insects from the Scale Insect Family Diaspididae. *Insects* **2012**, *3*, 262–269. [CrossRef] [PubMed]
11. Gruwell, M.E.; Morse, G.E.; Normark, B.B. Phylogenetic Congruence of Armored Scale Insects (Hemiptera: Diaspididae) and Their Primary Endosymbionts from the Phylum Bacteroidetes. *Mol. Phylogenetics Evol.* **2007**, *44*, 267–280. [CrossRef]
12. Sabree, Z.L.; Huang, C.Y.; Okusu, A.; Moran, N.A.; Normark, B.B. The Nutrient Supplying Capabilities of *Uzinura*, an Endosymbiont of Armoured Scale Insects. *Environ. Microbiol.* **2013**, *15*, 1988–1999. [CrossRef]
13. Szklarzewicz, T.; Michalik, A.; Michalik, K. The Diversity of Symbiotic Systems in Scale Insects. In *Symbiosis: Cellular, Molecular, Medical and Evolutionary Aspects*; Kloc, M., Ed.; Springer International Publishing: Cham, Switzerland, 2020; pp. 469–495, ISBN 978-3-030-51849-3.
14. van den Bosch, T.J.M.; Welte, C.U. Detoxifying Symbionts in Agriculturally Important Pest Insects. *Microb. Biotechnol.* **2017**, *10*, 531–540. [CrossRef]
15. Silva, D.P.; Epstein, H.E.; Vega Thurber, R.L. Best Practices for Generating and Analyzing 16S RRNA Amplicon Data to Track Coral Microbiome Dynamics. *Front. Microbiol.* **2023**, *13*, 1007877. [CrossRef]
16. Straub, D.; Blackwell, N.; Langarica-Fuentes, A.; Peltzer, A.; Nahnsen, S.; Kleindienst, S. Interpretations of Environmental Microbial Community Studies Are Biased by the Selected 16S RRNA (Gene) Amplicon Sequencing Pipeline. *Front. Microbiol.* **2020**, *11*, 550420. [CrossRef]
17. Agredo-Gomez, A.D.; Molano-Molano, J.A.; Portela-Patiño, M.C.; Rodríguez-Páez, J.E. Use of ZnO Nanoparticles as a Pesticide: In Vitro Evaluation of Their Effect on the Phytophagous Puto Barberi (Mealybug). *Nano-Struct. Nano-Objects* **2024**, *37*, 101095. [CrossRef]
18. Ullah, M.S.; Sharif, M.M.H.; Dey, A.; Haque, M.M.; Rashed, M.T.N.N. Capacity of Detergent to Increase the Efficacy of Bio-Pesticides and Chemical Pesticides against Giant Mealybug, Drosicha Mangiferae (Homoptera: Pseudococcidae). *Int. J. Trop. Insect Sci.* **2024**, *44*, 2889–2896. [CrossRef]
19. El Aalaoui, M.; Sbaghi, M. Population Fluctuations, Diversity and Effectiveness of Natural Enemies Associated with the Cactus Scale *Diaspis Echinocacti* (Bouché) (Hemiptera: Diaspididae) in Morocco. *Phytoparasitica* **2023**, *51*, 1059–1072. [CrossRef]
20. Callahan, B.J.; McMurdie, P.J.; Rosen, M.J.; Han, A.W.; Johnson, A.J.A.; Holmes, S.P. DADA2: High-Resolution Sample Inference from Illumina Amplicon Data. *Nat. Methods* **2016**, *13*, 581–583. [CrossRef] [PubMed]
21. Bolyen, E.; Rideout, J.R.; Dillon, M.R.; Bokulich, N.A.; Abnet, C.C.; Al-Ghalith, G.A.; Alexander, H.; Alm, E.J.; Arumugam, M.; Asnicar, F.; et al. Reproducible, Interactive, Scalable and Extensible Microbiome Data Science Using QIIME 2. *Nat. Biotechnol.* **2019**, *37*, 852–857. [CrossRef]
22. Douglas, A.E. Multiorganismal Insects: Diversity and Function of Resident Microorganisms. *Annu. Rev. Entomol.* **2015**, *60*, 17–34. [CrossRef]
23. Chen, B.; Teh, B.-S.; Sun, C.; Hu, S.; Lu, X.; Boland, W.; Shao, Y. Biodiversity and Activity of the Gut Microbiota across the Life History of the Insect Herbivore Spodoptera Littoralis. *Sci. Rep.* **2016**, *6*, 29505. [CrossRef] [PubMed]
24. Yun, J.-H.; Roh, S.W.; Whon, T.W.; Jung, M.-J.; Kim, M.-S.; Park, D.-S.; Yoon, C.; Nam, Y.-D.; Kim, Y.-J.; Choi, J.-H.; et al. Insect Gut Bacterial Diversity Determined by Environmental Habitat, Diet, Developmental Stage, and Phylogeny of Host. *Appl. Environ. Microbiol.* **2014**, *80*, 5254–5264. [CrossRef]
25. Caspi, R.; Billington, R.; Fulcher, C.A.; Keseler, I.M.; Kothari, A.; Krummenacker, M.; Latendresse, M.; Midford, P.E.; Ong, Q.; Ong, W.K.; et al. The MetaCyc Database of Metabolic Pathways and Enzymes. *Nucleic Acids Res.* **2018**, *46*, D633–D639. [CrossRef]
26. Nicol, M.; Alexandre, S.; Luizet, J.-B.; Skogman, M.; Jouenne, T.; Salcedo, S.P.; Dé, E. Unsaturated Fatty Acids Affect Quorum Sensing Communication System and Inhibit Motility and Biofilm Formation of Acinetobacter Baumannii. *Int. J. Mol. Sci.* **2018**, *19*, 214. [CrossRef]

27. Heath, R.J.; Rock, C.O. Roles of the FabA and FabZ β-Hydroxyacyl-Acyl Carrier Protein Dehydratases in Escherichia Coli Fatty Acid Biosynthesis. *J. Biol. Chem.* **1996**, *271*, 27795–27801. [CrossRef]
28. Huerta-García, A.; Álvarez-Cervantes, J. The Gut Microbiota of Insects: A Potential Source of Bacteria and Metabolites. *Int. J. Trop. Insect Sci.* **2024**, *44*, 13–30. [CrossRef]
29. Beltrame, W.S.; Daquila, B.V.; Caleffe, R.R.T.; dos Santos Wolff, V.R.; da Silva Santos, É.; da Silva Machado, M.D.F.P.; Conte, H.; Ruvolo-Takasusuki, M.C.C. Gall Formation in Cereus Sp. Infected with *Diaspis Echinocacti* Bouché, 1833 (Hemiptera: Diaspididae). *Flora* **2022**, *290*, 152042. [CrossRef]
30. El Fakhouri, K.; Ramdani, C.; Aasfar, A.; Boulamtat, R.; Sijilmassi, B.; El Bouhssini, M.; Kadmiri, I.M. Isolation, Identification and Pathogenicity of Local Entomopathogenic Bacteria as Biological Control Agents against the Wild Cochineal Dactylopius Opuntiae (Cockerell) on Cactus Pear in Morocco. *Sci. Rep.* **2023**, *13*, 21647. [CrossRef] [PubMed]
31. Callejas-Chavero, A.; Martínez-Hernández, D.; Flores-Martínez, A.; Moncada-Orellana, A.; Diaz-Quiñones, Y.; Vargas-Mendoza, C.F. Herbivory in Cacti: Fitness Effects of Two Herbivores, One Tending Ant on Myrtillocactus Geometrizans (Cactaceae). In *Evolutionary Ecology of Plant-Herbivore Interaction*; Springer International Publishing: Cham, Switzerland, 2020; pp. 109–134.
32. Husnik, F.; McCutcheon, J.P. Repeated Replacement of an Intrabacterial Symbiont in the Tripartite Nested Mealybug Symbiosis. *Proc. Natl. Acad. Sci. USA* **2016**, *113*, E5416–E5424. [CrossRef]
33. Yuan, Z.; Liu, F.; Yuan, Y.; Pan, H. Structural Composition and Diversity of Bacterial Communities in High- and Low-Yielding Moso Bamboo Forests. *Front. Biosci.-Landmark* **2023**, *28*, 290. [CrossRef]
34. Kondo, T.; Watson, G.W. *Encyclopedia of Scale Insect Pests*; CAB International: Wallingford, UK, 2022; ISBN 978-1-80062-065-0.
35. Rosenblueth, M.; Martínez-Romero, J.; Ramírez-Puebla, S.T.; Vera-Ponce de León, A.; Rosas-Pérez, T.; Bustamante-Brito, R.; Rincón-Rosales, R.; Martínez-Romero, E. Microorganismos Endosimbiontes de Insectos Escama. *TIP Rev. Espec. Cienc. Quím.-Biol.* **2018**, *21*, 53–69.
36. El Aalaoui, M.; Sbaghi, M. Ecological Dynamics of Phenacoccus Solenopsis Tinsley (Hemiptera: Pseudococcidae) and Its Parasitoids in Varied Host Environments. *Phytoparasitica* **2024**, *52*, 44. [CrossRef]
37. Yasika, Y.; Shivakumar, M.S. A Comprehensive Account of Functional Role of Insect Gut Microbiome in Insect Orders. *J. Nat. Pestic. Res.* **2025**, *11*, 100110. [CrossRef]
38. Wang, S.; Wang, L.; Fan, X.; Yu, C.; Feng, L.; Yi, L. An Insight into Diversity and Functionalities of Gut Microbiota in Insects. *Curr. Microbiol.* **2020**, *77*, 1976–1986. [CrossRef]
39. Lu, Q.-C.; Yu, J.-M.; Liu, H.-L.; Wu, X.-L.; Wei, S.-J.; Lei, M.; Cai, P.; He, H.-G.; Pu, D.-Q. Stable Composition of Gut Microbiome in the Asian Ladybeetle Coccinella Septempunctata Reared on Natural and Artificial Diets. *Sci. Rep.* **2024**, *14*, 71. [CrossRef]
40. Kumar, D.; Sun, Z.; Cao, G.; Xue, R.; Hu, X.; Gong, C. Bombyx Mori Bidensovirus Infection Alters the Intestinal Microflora of Fifth Instar Silkworm (Bombyx Mori) Larvae. *J. Invertebr. Pathol.* **2019**, *163*, 48–63. [CrossRef]
41. Ranjith, M.T.; ManiChellappan; Harish, E.R.; Girija, D.; Nazeem, P.A. Bacterial Communities Associated with the Gut of Tomato Fruit Borer, Helicoverpa Armigera (Hübner) (Lepidoptera: Noctuidae) Based on Illumina Next-Generation Sequencing. *J. Asia Pac. Entomol.* **2016**, *19*, 333–340. [CrossRef]
42. Ali, I.; Zhang, S.; Luo, J.-Y.; Wang, C.-Y.; Lv, L.-M.; Cui, J.-J. Artificial Diet Development and Its Effect on the Reproductive Performances of Propylea Japonica and Harmonia Axyridis. *J. Asia Pac. Entomol.* **2016**, *19*, 289–293. [CrossRef]
43. Pongen, Y.L.; Thirumurugan, D.; Ramasubburayan, R.; Prakash, S. Harnessing Actinobacteria Potential for Cancer Prevention and Treatment. *Microb. Pathog.* **2023**, *183*, 106324. [CrossRef]
44. Salgado, J.F.M.; Hervé, V.; Vera, M.A.G.; Tokuda, G.; Brune, A. Unveiling Lignocellulolytic Potential: A Genomic Exploration of Bacterial Lineages within the Termite Gut. *Microbiome* **2024**, *12*, 201. [CrossRef]
45. Chen, L.; He, Z.; Zhang, D.; Zhao, F.; Zhang, Y.; Ding, R. The Role of Gut Microbiota at Different Developmental Stages in the Adaptation of the Etiella Zinckenella to a Plant Host. *Sci. Rep.* **2025**, *15*, 4971. [CrossRef]
46. Moran, N.A.; McCutcheon, J.P.; Nakabachi, A. Genomics and Evolution of Heritable Bacterial Symbionts. *Annu. Rev. Genet.* **2008**, *42*, 165–190. [CrossRef]
47. Wikström, B.; Huemer, P.; Mutanen, M.; Tyllinen, J.; Kaila, L. Pyralis Cardinalis, a Charismatic New Species Related to P. Regalis [Denis & Schiffermüller], 1775, First Recognized in Finland (Lepidoptera, Pyralidae). *Nota Lepidopterol.* **2020**, *43*, 337–364. [CrossRef]
48. Zhang, Y.-M.; Rock, C.O. Membrane Lipid Homeostasis in Bacteria. *Nat. Rev. Microbiol.* **2008**, *6*, 222–233. [CrossRef]
49. Kinjo, Y.; Bourguignon, T.; Hongoh, Y.; Lo, N.; Tokuda, G.; Ohkuma, M. Coevolution of Metabolic Pathways in Blattodea and Their *Blattabacterium* Endosymbionts, and Comparisons with Other Insect-Bacteria Symbioses. *Microbiol. Spectr.* **2022**, *10*, e02779-22. [CrossRef]

biology

MDPI

Article

Assessment of Potential Toxic Effects of RNAi-Based Transgenic Cotton on the Non-Target Predator *Harmonia axyridis*

Haiqin Yao [1,†], Haonan Xu [2,†], Jun Yang [1,*] and Weihua Ma [2]

1 College of Food Science and Technology, Wuhan Business University, Wuhan 430056, China; yaohq@wbu.edu.cn
2 Hubei Insect Resources Utilization and Sustainable Pest Management Key Laboratory, College of Plant Science and Technology, Huazhong Agricultural University, Wuhan 430070, China
* Correspondence: yjun202201@163.com; Tel.: +86-13397175595
† These authors contributed equally to this work.

Simple Summary: The deployment of genetically engineered (GE) crops, particularly RNA interference (RNAi) crops that express double-stranded RNA, has raised biosafety concerns among regulatory authorities, the scientific community, and the public. Such crops may present potential risks to public health, agricultural sustainability, and biodiversity. Natural enemies of insect pests, including the lady beetle *Harmonia axyridis*, play a crucial role in pest regulation and maintaining ecosystem balance. This study assessed the potential effects of RNAi cotton targeting the pest *Adelphocoris suturalis* on *H. axyridis*. Both laboratory and field experiments demonstrated no significant adverse impacts on the beetle's growth, development, predatory efficacy, or reproductive capacity. Furthermore, the RNAi construct was not effectively transmitted through the food chain. These results support the environmental safety of RNAi cotton within the plant–pest–natural enemy ecological framework. This contributes to the advancement of sustainable agroecosystems, the conservation of biodiversity, and the production of safer food supplies, ultimately benefiting agriculture and society by enhancing crop protection while minimizing ecological risks.

Abstract: Although traditional genetically modified (GM) cotton has reduced lepidopteran pests, secondary pests such as *Adelphocoris suturalis* Jakovlev (Hemiptera: Miridae) have become increasingly problematic. RNA interference (RNAi)-based insect-resistant plants offer a promising alternative, but their potential ecological risks, particularly within the plant–pest–natural enemy framework, require comprehensive evaluation. As a key natural enemy in cotton agroecosystems, *Harmonia axyridis* Pallas (Coleoptera: Coccinellidae) plays a vital role in regulating pest populations and thus serves as an important non-target insect for environmental risk assessment. In this study, ds*AsFAR* transgenic cotton and the non-target insect *H. axyridis* were used to evaluate potential ecological risks. Based on the sequences of the *HaFAR*, *AsFAR*, and *GFP* genes, ds*AsFAR*, ds*HaFAR*, and ds*GFP* were synthesized in vitro and incorporated into artificial diets fed to *H. axyridis*, along with a ddH_2O control group. No significant differences were observed among the treatments across various developmental stages. Furthermore, *H. axyridis* was fed aphids that had been reared on either transgenic or non-transgenic cotton plants cultivated in a greenhouse. The findings indicated no significant differences in the growth, development, predatory ability, or fecundity of *H. axyridis*. These results suggest that transgenic ds*AsFAR* cotton targeting *A. suturalis* poses no detectable adverse effects on *H. axyridis*, supporting the environmental safety of RNAi crops within a plant–pest–natural enemy ecological interaction framework. Moreover, the ds*AsFAR* expressed in transgenic cotton was not effectively transferred through the food chain to affect homologous gene expression in *H. axyridis*.

Keywords: risks assessment; RNAi; *Adelphocoris suturalis*; plant–pest–natural enemy; ds*AsFAR*

1. Introduction

RNA interference (RNAi) is an evolutionarily conserved post-transcriptional gene silencing (PTGS) phenomenon in which the introduction of dsRNA leads to the degradation of specific mRNA and inhibits the normal expression of target genes [1,2]. Matzke was the first to report the phenomenon of gene silencing caused by tobacco co-transfection [3]. With the rapid development of modern molecular technology, RNAi is increasingly being recognized as one of the leading approaches for managing major agricultural pests due to its environmentally friendly, efficient, and target-specific characteristics [4]. The quality of various crops has been improved by using RNAi technology, for example, via interfering with the function of fatty acid dehydrogenase, which significantly increases the proportion of oleic acid in rapeseed [5]. Including RNAi crops, insect-resistant genetically engineered (IRGE) crops that use RNAi have been released to the market [6]. In order to gain a deeper understanding of their inherent risks to the ecological environment, most countries that have approved the environmental release of RNAi crops have begun to develop regulatory frameworks to evaluate the potential hazards of using this new biotechnology [7–9]. However, there has been no standard framework for assessing the environmental risk of RNAi crops in China up to now.

Cotton is widely cultivated as a cash crop worldwide, but its safe production is threatened by various factors, such as agricultural pests, diseases, drought, soil salinization, and so on [10,11]. In view of the above reasons, genetically modified technology has profound implications for cotton cultivation. With the continuous development of molecular biotechnology, the planting area of transgenic cotton has reached more than 90% in China [12]. The popularization and cultivation of traditional genetically modified (GM) insect-resistant cotton has significantly reduced the population density of lepidopteran pests such as cotton bollworms in the field. However, due to large-scale monoculture and a reduction in the use of chemical pesticides, minor pests such as mites and aphids have gradually become rampant and become major pests. Among them, *A. suturalis* and *Apolygus lucorum* Meyer—Dür (Heteroptera: Miridae) can cause serious damage to cotton production and quality. Transgenic ds*AsFAR* cotton can inhibit the fecundity of *A. suturalis* by silencing its *FAR* gene expression. This type of dsRNA transgenic cotton can produce significant control effects on field populations of *A. suturalis* and has great application prospects in production. However, its ecological safety within the plant–pest–natural enemy framework has not yet been fully evaluated. *H. axyridis*, a common natural enemy insect, plays a crucial ecological role in cotton field ecosystems by preying on various pests. As a representative species in the plant–pest–natural enemy trophic system, *H. axyridis* is ecologically significant for maintaining pest population balance. Moreover, as a member of the Coleoptera order, *H. axyridis* has been reported to exhibit higher sensitivity to dsRNA compared to insects from other insect orders, making it an ideal test organism for evaluating the biosafety of transgenic dsRNA cotton in non-target beneficial insects [13].

A previous study conducted by our team found that the direct feeding of ds*AsFAR* to *H. axyridis* had no significant impact on their own growth, development, reproduction, or other physiological processes, and that GM cotton does not affect physiological indicators of *H. axyridis* through the food chain [14]. However, functional genomics research shows that RNAi often induces off-target effects in cells or organisms, which is a major drawback of this technology [15–17]. The off-target effects of RNAi on non-target organisms may be sublethal, and such effects may not be readily detectable through biological assays alone. Additionally, organisms possess a vast number of genes, complicating the assessment of these effects. Omics analysis techniques [18] offer a comprehensive approach to detecting the off-target effects of RNAi transgenic crops by examining global transcriptome changes. We employed qRT-PCR to investigate the relationship between dsRNA uptake efficiency and the duration of feeding. Additionally, transcriptome sequencing technology combined with quantitative analysis verification was utilized to assess the off-target effect of dsRNA. Furthermore, to anticipate the potential impact of off-target effects from transgenic dsRNA crops, this study employed RNAi technology to predict the possible consequences for non-target organisms. This approach provides a useful reference for evaluating the safety of non-target organisms consuming transgenic dsRNA crops over multiple generations under field conditions.

2. Materials and Methods

2.1. General Procedures

2.1.1. Insect Rearing

The initial population of *H. axyridis* was kindly provided by Xingmiao Zhou (Huazhong Agricultural University, Wuhan, China). The beetles were reared for several generations in an artificial climate chamber maintained at 25 ± 2 °C, 65 ± 5% relative humidity, and a 16 h light/8 h dark photoperiod. The insects were fed *Acyrthosiphon pisum* Harris (Hemiptera: Aphididae) reared on *Vicia faba* L. (Fabales: Fabaceae). To minimize cannibalism, larvae were placed individually in 90 mm Petri dishes containing moistened tissue paper. Artificial diet was provided daily, and larval development was monitored until adult emergence. The artificial diet was prepared as follows: (1) A dry mixture was made from pig liver powder, yeast, sucrose, linseed oil, olive oil, and β-carotene, thoroughly stirred with distilled water. (2) In parallel, agar and honey were dissolved in distilled water with heating. (3) The heated solution was boiled several times and added to the first mixture, then stirred to homogeneity; (4) The final mixture was dispensed into 24-well plates and stored at 4 °C after cooling below 50 °C.

2.1.2. Cotton Cultivation

Seeds of transgenic ds*AsFAR* cotton and its non-transgenic parental cultivar (*Gossypium hirsutum* cv. Jin668) were obtained from Prof. Shuangxia Jin (National Key Laboratory of Crop Genetic Improvement, Huazhong Agricultural University). The ds*AsFAR* line expresses a 432 bp double-stranded RNA targeting the *AsFAR* gene, which was introduced via Agrobacterium-mediated transformation using the pHellsgate4 RNAi vector. Both lines were planted in pesticide-free experimental plots (10.5 m × 4.5 m) at Huazhong Agricultural University.

2.2. Experimental Procedures

2.2.1. Trophic Transfer Assay

This assay was designed to evaluate the effect of transgenic cotton on *H. axyridis* via a tritrophic pathway. At the four-leaf stage, *A. pisum* aphids were transferred onto transgenic or control cotton plants and reared for three generations to ensure sufficient dsRNA uptake. Second-instar *H. axyridis* larvae were then introduced into Petri dishes (Φ = 90 mm) containing a fresh cotton leaf and aphid larvae. Life-table parameters—including survival, developmental time, and fecundity—were recorded. Each treatment included three replicates with 20 larvae per replicate.

2.2.2. Artificial Diet Exposure Assay

This direct RNAi assay aimed to determine the effect of orally delivered dsRNA on *H. axyridis*. Artificial diets were mixed with ds*AsFAR*, ds*HaFAR* (positive control), ds*GFP* (negative control), or ddH_2O. Each dsRNA was provided at a concentration 10-fold higher than that expressed in cotton leaves (7 μg/g). One newly molted 4th instar larva (<24 h) was placed in a Petri dish containing the treated diet. The diet was refreshed daily. Upon adult emergence, 10 individuals (6 females and 4 males) were housed in a 12 cm diameter, 10 cm high cylinder with *A. pisum* and three to four branches of *Jasminum nudiflorum* Lindl. (Oleaceae: Jasminum) as an oviposition substrate. Life-history traits—such as adult longevity, fecundity, and survival—were monitored. Each treatment was conducted with three replicates of 20 individuals. To ensure the reliability and validity of the experimental data, and to minimize potential confounding factors, the mortality rate of the control groups (artificial diets mixed with ds*GFP* or ddH_2O) were kept below 10%.

2.3. Synthesis of dsRNAs

Pairwise sequence alignment was carried out to obtain the *HaFAR* gene sequence most homologous to *AsFAR*, which is expressed in ds*AsFAR* transgenic cotton (49.55%) from our local *H. axyridis* transcriptome. Subsequently, ds*AsFAR*, ds*HaFAR*, and ds*GFP* [19] were amplified by PCR using primers containing the T7 promoter (Table S1). The dsRNAs were synthesized using a dsRNA Synthesis Kit (Thermo Fisher Scientific, CA, USA).

2.4. Quantitative Real-Time PCR (qRT-PCR)

Following the manufacturer's instructions, total RNA was extracted with the RNAiso reagent (Takara, Kyoto, Japan). cDNA was prepared using a reverse transcription kit (Takara, Kyoto, Japan). The 10 μL qRT-PCR reaction mixture consisted of 5 μL of SYBR solution (Takara, Kyoto, Japan), 0.8 μL of specific primers (Table S1), 2.2 μL of double-distilled water, and 2 μL of cDNA. The reaction mixture was placed in a 96-well microplate. The qRT-PCR was performed on the Bio-Rad Detection iQ2 System (Bio-Rad, Hercules, CA, USA). The qRT-PCR program was 95 °C for 30 s, followed by 40 cycles of 95 °C for 5 s and 60 °C for 30 s. The *H. axyridis 18S* gene was used as an endogenous reference gene for the run. The $2^{-\Delta\Delta Ct}$ method was used to calculate the relative transcript levels of the corresponding genes.

2.5. Off-Target Effects and Transcriptome Analysis

The samples collected above were also used for RNA-Seq on an Illumina platform at MetWare (Wuhan, China). The raw reads were filtered, and the clean reads were de novo assembled into unigenes using Trinity. Transcript abundance was then estimated using RSEM. Differential expression analysis was performed using DESeq2. The screening of differentially expressed genes (DEGs) was based on $\log_2$ (fold change) and false discovery rate (FDR). Genes with $\log_2$ (fold change) > 1 and FDR < 0.05 were considered DEGs. Several randomly selected DEGs from each group were validated using qRT-PCR. The off-target effects of dsRNA were assessed by analyzing the correlation between the number of matched base pairs and their fold changes for each gene containing continuously matched base pairs, using Perl scripts. Furthermore, two additional sets of non-target genes were also considered. The first one was the homologous genes of the target genes, which were identified using BLASTN (E-value $< 1 \times 10^{-10}$). The second one was genes in the same pathway as the target genes. In addition, the transcriptome was evaluated using the Shannon entropy count, which was calculated using R script. Defining transcriptomic changes as Shannon entropy allowed transcriptome variation to be displayed as a separate metric [20]. The bioinformatic analysis was adopted from our previous study [21]. The enrichment analysis was performed based on the hypergeometric test. For the Kyoto Encyclopedia of Genes and Genomes (KEGG) analysis, the hypergeometric distribution test was performed at the pathway level; for Gene Ontology (GO), it was performed based on the GO term. Data visualization was performed using R (version 4.2.0).

2.6. Statistical Analysis

GraphPad Prism 10.1.2 (Dotmatics, Boston, MA, USA) was used for the statistical analysis of the data. Data from the biological assays were analyzed using one-way ANOVA followed by Tukey's multiple comparison tests, and data produced by qRT-PCR were analyzed using Student's *t*-test.

3. Results

3.1. Uptake Efficiency of dsRNA in H. axyridis

The expression level of dsRNA in ds*AsFAR* transgenic cotton is 0.7 μg/g. The dsRNA was added to the artificial diet, resulting in a final concentration of 7 μg/g (10-fold higher than in GM cotton) for ds*AsFAR*, ds*HaFAR* (positive control), and ds*GFP* (negative control), respectively. Both the larvae and adults of *H. axyridis* were fed on all dsRNAs for 3 and 6 days and 3 days, respectively, and then the changes in *HaFAR* gene expression levels in *H. axyridis* were detected after 24 h of feeding. The results showed that ingesting ds*AsFAR* caused no significant interference in the *HaFAR* expression in *H. axyridis* larvae and adults (Figure 1A,B). In addition, the dsRNA uptake efficiency did not change significantly when the larvae were fed on dsRNA for different lengths (3 d and 6 d) compared with the positive control (ds*HaFAR*). However, the positive control exhibited higher RNAi efficiency in *H. axyridis* adults.

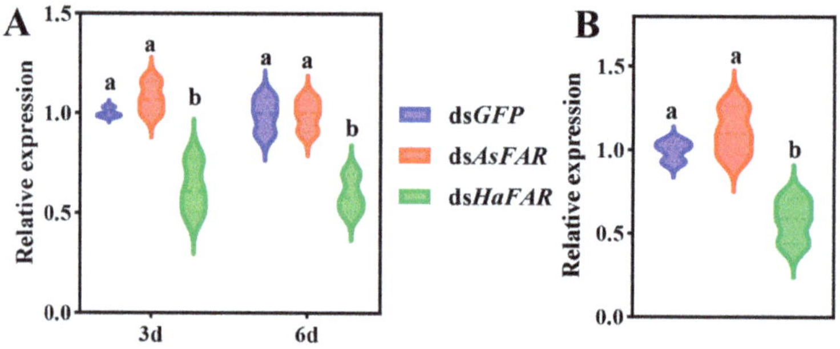

Figure 1. Changes in *FAR* gene expression in larvae and adults of *H. axyridis* after feeding on three dsRNAs. (**A**) Changes in *FAR* gene expression in larvae of *H. axyridis* after feeding on three dsRNAs for 3 d and 6 d (F_{3d} (2, 6) = 20.65, P_{3d} = 0.0020; F_{6d} (2, 6) = 19.83 P_{6d} = 0.0023). (**B**) Changes in *FAR* gene expression in adults of *H. axyridis* after feeding three dsRNAs for 3 d (F (2, 6) = 16.25, p = 0.0038). The values are represented as mean ± SD of three replicates. Error bars indicate the standard error of the mean. Different lowercase letters (a, b) indicated significant differences (p-value < 0.05). p-Values are based on multiple comparisons that were performed using one-way ANOVA followed by Tukey's multiple comparison tests.

3.2. Effect of Uptake dsRNA on the Life-Table Parameters of H. axyridis

Even though a 10-fold (7 μg/g) high dosage was applied in the feeding test, no significant changes were observed in different developmental stages of *H. axyridis* that were separately treated with dsRNAs (ds*GFP*, ds*AsFAR*, and ds*HaFAR*) and H_2O. The developmental durations of *H. axyridis* at each instar stage were relatively consistent (Figure 2A). All the survival rates of *H. axyridis* with dsRNA treatments were above 95%, but there was no significant difference among them (Figure 2B). In addition, the pupal weight (Figure 2C), emergence rate (Figure 2D), adult weight (Figure 2E), number of predations by females (Figure 2F), number of predations by males (Figure 2G), egg production (Figure 2H), and hatching rates (Figure 2I) for populations from all treatments showed no significant differences. Furthermore, the *HaFAR* gene expression level in the ds*HaFAR* treatment was significantly decreased compared with other treatments (p-value < 0.05, Figure 2J), indicating that *H. axyridis* can acquire exogenous dsRNA through feeding. Overall, these results demonstrate that ds*AsFAR* had no significant effect on the survival, development, predation ability, fecundity, or *HaFAR* expression of *H. axyridis* at any stage.

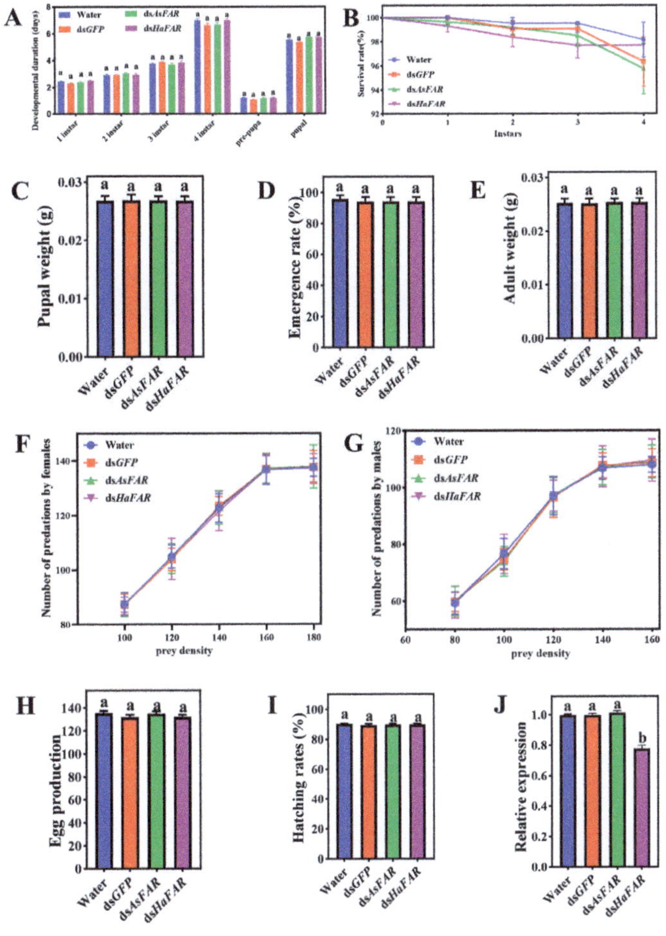

Figure 2. Evaluation of RNA interference of *FAR* in *H. axyridis*. Duration of the pre-imaginal stages (**A**), survival rates (**B**), pupal weight ($F_{1\ instar}$ (3, 282) = 1.85, $P_{1\ instar}$ = 0.14; $F_{2\ instar}$ (3, 278) = 0.6440, $P_{2\ instar}$ = 0.5873; $F_{3\ instar}$ (3, 276) = 0.67, $P_{4\ instar}$ = 0.5719; $F_{4\ instar}$ (3, 268) = 1.556, $P_{pre\text{-}pura}$ = 0.2004; $F_{pre\text{-}pura}$ (3, 268) = 1.533, P_{pupal} = 0.2063 F (3, 253) = 1.551, p = 0.2019) (**C**), emergence rate (F (3, 268) = 0.28, p = 0.84) (**D**), adult weight (F (3, 268) = 0.09, p = 0.97) (**E**), number of predations by females (**F**), number of predations by males (**G**), egg production (F (3, 56) = 0.77, p = 0.51) (**H**), and hatching rates (F (3, 268) = 0.28, p = 0.84) (**I**) for populations from all treatments. (**J**) Changes in *FAR* expression after feeding dsRNA-mixed artificial diets in *H. axyridis* (F (3, 8) = 209.14, $p = 6.15 \times 10^{-8}$). The values are represented as mean ± SD of three replicates. Error bars indicate the standard error of the mean. Different lowercase letters (a, b) indicate significant differences (p-value < 0.05). p-Values are based on multiple comparisons that were performed using one-way ANOVA followed by Tukey's multiple comparison tests.

3.3. *The Vital Activities of H. axyridis Were Not Affected by the Trophic Transfer Assay*

The ds*AsFAR* transgenic cotton and its parent Jin668 were planted in the greenhouse, and the cotton aphids were transferred to different types of cotton plants for propagation and used to feed *H. axyridis*. The results showed that no significant difference was observed in the development duration (Figure 3A), survival rate (Figure 3B), pupal and adult weight (Figure 3C,E), number of predations by females (Figure 3F), number of predations by males (Figure 3G), egg production (Figure 3H), or hatching rate (Figure 3I) of *H. axyridis* when compared with the control treatment, indicating that there was no adverse effect of ds*AsFAR* transgenic cotton on the survival and development of *H. axyridis*. Adults of *H. axyridis* that fed on aphids reared on different types of cotton plants were collected and analyzed by qPCR. The results showed that the ds*AsFAR* transgenic cotton had no adverse effect on the homologous gene expression of *H. axyridis* through the trophic transfer assay (Figure 3J).

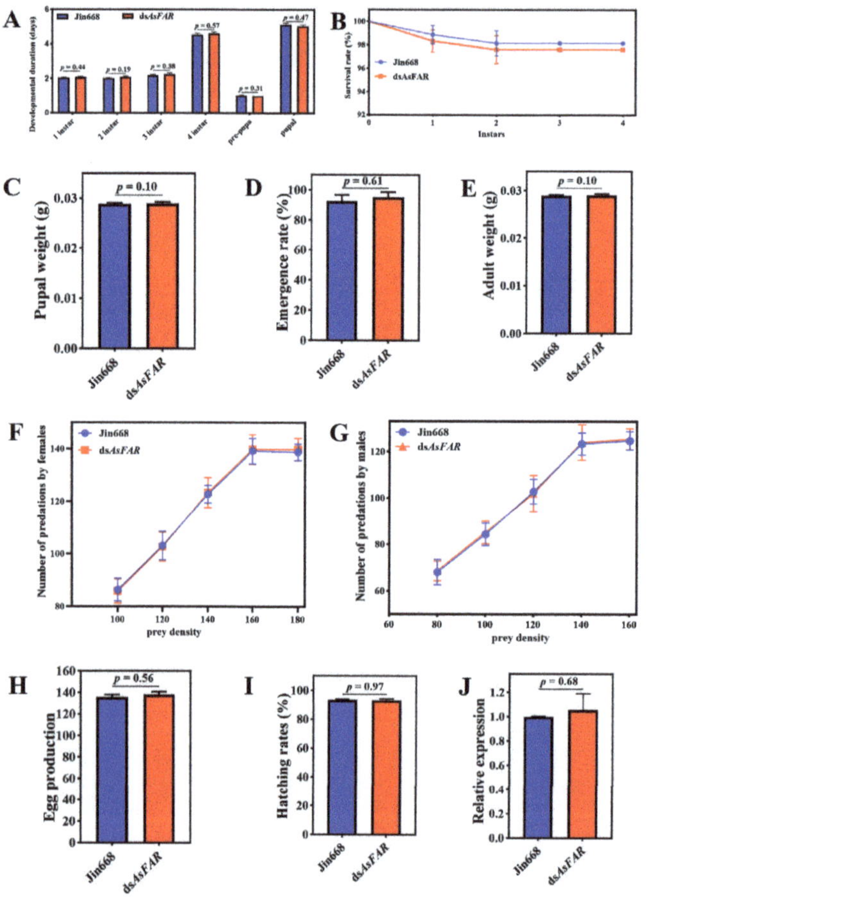

Figure 3. Life-table parameters of *H. axyridis* through the trophic transfer assay. Duration of the pre-imaginal stages ($df_{1\ instar}$ = 82, $df_{2\ instar}$ = 80, $df_{3\ instar}$ = 80, $df_{4\ instar}$ = 80, $df_{pre\text{-}pura}$ = 80, df_{pupal} = 75) (**A**), survival rates (**B**), pupal weight (df = 80) (**C**), emergence rate (df = 80) (**D**), adult weight (df = 80) (**E**), number of predations by females (**F**), number of predations by males (**G**), egg production (df = 28) (**H**), and hatching rates (df = 28) (**I**) of *H. axyridis* after feeding on aphids that were reared on ds*AsFAR* and jin668 cottons (df = 28). (**J**) Changes in *FAR* expression after feeding on Jin668 and ds*AsFAR* cotton in *H. axyridis* (df = 4). The values are represented as mean ± SD of three replicates. Error bars indicate the standard error of the mean. Each group was compared by Student's *t*-test.

3.4. Off-Target Effects on Non-Homologous Genes Were Induced by dsAsFAR in H. axyridi

According to the results of *H. axyridis* consumption of dsRNA, samples of *H. axyridis* that fed on ds*AsFAR* and ds*GFP* for 4 days at the larval and adult stages were sent to transcriptome sequencing. The results showed that exposure to ds*AsFAR* induced down-regulation of two genes in larvae and eight genes in adults, compared to ds*GFP*, but not in the *HaFAR* gene family (Figure 4A, Tables S2 and S3). We randomly selected eight differentially expressed genes (DEGs) for qRT-PCR assays to validate the transcriptome data. The expression profiles of all selected genes were consistent with the transcriptome results in both larvae and adults (Figure 4B,C). KEGG enrichment analysis of DEGs showed that DEGs in *dsAsFAR*-treated larvae were only significantly enriched in the non-homologous end-joining pathway; however, DEGs in adults were significantly enriched in protein processing in the endoplasmic reticulum, glycosaminoglycan degradation, neutrophil extracellular trap formation, and toxoplasmosis (Figure S1). The results of GO enrichment analysis were consistent with those of KEGG analysis, with both showing no correlation with lipid or wax synthesis (Figure S2). Therefore, we speculate that ds*AsFAR* may induce the off-target effects of the non-homologous genes in the *H. axyridis*.

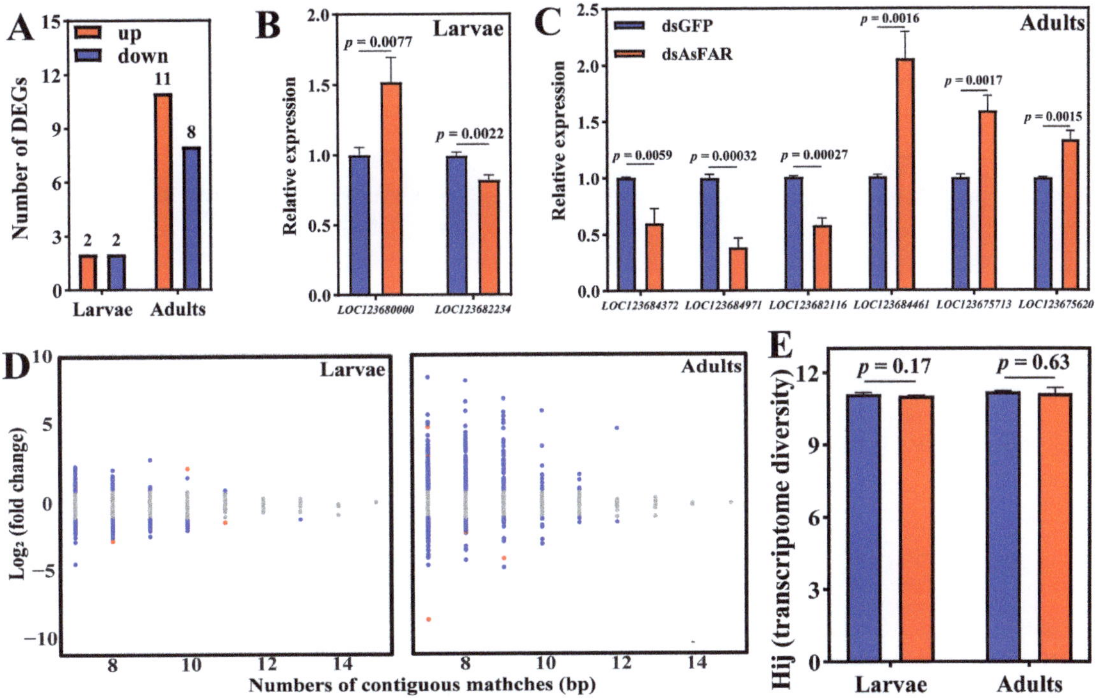

Figure 4. Transcriptome entropy for evaluating siRNA off-effects. Quantitative count of DEGs (**A**) for transcriptome sequencing and qRT-PCR validation for *O. similis* larvae (df = 4) (**B**) and adults (df = 4) (**C**). (**D**) The number of base pairs matched is linked with their fold changes for each continuous matched gene. Red, upregulated genes; blue, downregulated genes; gray, no significant genes. (**E**) The effect of siRNA feed on the Shannon transcriptome entropy (df = 4). The values are represented as mean ± SD of three replicates. Error bars indicate the standard error of the mean. Each group was compared by Student's *t*-test.

3.5. The Evaluation of Off-Target Effects and Transcriptome Homeostasis

The interference fragment of ds*AsFAR* we selected did not find a homologous fragment in *H. axyridis* (E-value < 1×10^{-5}). Through analysis of the genes containing continuous matches with dsRNA, the results showed that off-target effects were not correlated with the number of consecutive matching base pairs. Most DEGs of *H. axyridis* that fed on ds*AsFAR* have less than 10 consecutive matching base pairs. In addition, we also found that 99.90% to 100% of *H. axyridis* genes containing continuously matched regions with ds*AsFAR* were not affected (Table 1). The correlation between the number of matches and their fold changes showed that the highest expression changes were usually detected on genes with fewer consecutive matches (Figure 4D). This indicated that the causal relationships between the number of base matches and changes in their expression levels were diverse. Within the same gene function network, silencing of one gene may induce changes in the expressions of other genes in pathways that interact with level 1 genes (level 2 pathway in Table 2) and so on (level 3 pathway in Table 2).

Table 1. The silencing of the genes containing continuously matched regions with ds*AsFAR* in the *H. axyridis* transcriptome.

No. of Matched Base Pairs (bp)	Total No. of Genes	No. and Percentage of Up-Regulated Genes	No. and Percentage of Down-Regulated Genes	No. and Percentage of Non-Significant-Change Genes
7	5462	3 (0.05%)	3 (0.05%)	5456 (99.90%)
8	3722	2 (0.05%)	1 (0.03%)	3719 (99.92%)
9	1272	0 (0)	1 (0.08%)	1271 (99.92%)
10	325	0 (0)	0 (0)	325 (100%)
11	103	0 (0)	0 (0)	103 (100%)
12	21	0 (0)	0 (0)	21 (100%)
13	9	0 (0)	0 (0)	9 (100%)
14	3	0 (0)	0 (0)	3 (100%)
15	1	0 (0)	0 (0)	1 (100%)
Total	10,918	5 (0.05%)	5 (0.05%)	10,908 (99.90%)

Table 2. The silencing of the related KEGG genes after ds*AsAFR* ingestions in *H. axyridis*.

Transcriptome	KEGG Pathway Level	Total No. of Genes	No. and Percentage of Up-Regulated Genes	No. and Percentage of Down-Regulated Genes	No. and Percentage of Non-Significant Change Genes
Larvae	Level 2	2595	0 (0.00%)	0 (0.00%)	2595 (100.00%)
	Level 3	23,952	7 (0.03%)	3 (0.01%)	23,942 (99.96%)
	Total	26,547	7 (0.03%)	3 (0.01%)	26,537 (99.96%)
Adults	Level 2	2698	190 (7.04%)	6 (0.22%)	2502 (92.74%)
	Level 3	25,545	1261 (4.94%)	39 (0.25%)	24,245 (94.91%)
	Total	28,243	1451 (5.14%)	45 (0.16%)	26,747 (94.70%)

Shannon entropy of a transcriptome reflects the overall variability in gene expression and can be used to assess potential risks to fundamental biological processes that may arise from exposure to exogenous dsRNA. The results showed that the Shannon entropy values did not change significantly following ingestion of ds*AsFAR* in both larvae and adults (Figure 4E), which demonstrates that ds*AsFAR* has no adverse effect on the transcriptome homeostasis of *H. axyridis*.

4. Discussion

Traditional chemical pesticides have historically played a crucial role in contemporary agriculture by effectively managing insect pests, weeds, and plant pathogens, thereby significantly enhancing crop yields and contributing to global food security. Nonetheless, their extensive application has generated substantial ecological and environmental concerns. Conventional chemical pesticides frequently display broad-spectrum toxicity, which can adversely affect non-target insects and plants. Additionally, these substances have the potential to leach into or run off into adjacent water bodies and soil, thereby amplifying their ecological impact [22].

In contrast, GM technology provides new ideas for increasing crop yields and improving the quality of agricultural products. With the widespread application of IRGE crops and other novel biotechnologies in insect-pest management [19], rather than synthetic insecticides, many concerns about both their negative and positive effects have arisen in many countries. An effective and standardized risk evaluation framework for genetically engineered crops is necessary before their promotion and application. Such a framework not only helps to prevent unintended consequences to public health, agriculture, and conservation, but also facilitates the authentication and communication of the same biotechnology product among governmental regulators worldwide. The more popular IRGE crops are transgenic Bt insect-resistant crops, for which many studies on their environmental risks have been published so far. RNAi technology, as an efficient tool for interpreting sequence function and expression, is widely used in various fields [23], especially in gene function analysis and genetic breeding of plants. For example, transgenic materials with inserted dsRNA have been used to increase tobacco's resistance to insects [24] and to improve the resistance of cotton to cotton bollworm [25,26]. In addition, by integrating Bt toxin and RNAi, researchers have created transgenic cotton plants that target the juvenile hormone methyltransferase gene in cotton bollworm, effectively postponing insect pairing [27]. Additionally, RNAi has also been used to silence the potato beetle's highly specific ecdysone-related receptor genes, resulting in a mortality rate up to 80% and a significant reduction in larval weight and pupal weight [28]. Although significant progress has been made in researching these new genetically modified materials, there are still many potential environmental safety risks, including gene drift and off-target effects. Importantly, the off-target problem of RNAi has gradually emerged as RNAi research has deepened [16]. Hence, the off-target effect of dsRNA has become one of the important indicators for the safety evaluation of transgenic dsRNA plants [29,30]. However, there is no conclusive evidence regarding their possible environmental risks.

Natural enemies of insects play a crucial role in agricultural ecosystems [31]. Within the plant–pest–natural enemy framework, previous assessments of the ecological risks associated with insect-resistant IRGE crops have predominantly focused on evaluating their potential effects on the development and survival of beneficial arthropods, especially through trophic transfer assays. For example, Chen et al. reported that miRNA-mediated transgenic rice targeting the stem borer had no significant effect on *Apis mellifera* Linnaeus (Hymenoptera: Apidae), an essential pollinator [32]. Toxicity evaluations based on trophic transfer within the food chain are comparable to the whole-food animal tests frequently utilized in regulatory assessments of the potential risks posed by IRGE crops to human health. Consequently, it is posited that food chain-based testing constitutes one of the most suitable approaches for assessing the exposure pathways affecting natural enemies within the plant–pest–natural enemy system. Nonetheless, alternative approaches and validation strategies should also be explored prior to its routine implementation. In parallel, animal feeding tests facilitate the administration of significantly elevated doses of specific insecticidal components, thereby incorporating a conservative or worst-case exposure scenario into the risk assessment framework. For instance, the U.S. Environmental Protection Agency recommends employing exposure levels that are 10 times or more than typical field concentrations in laboratory risk assessments [33]. This approach ensures a high degree of safety assurance. Nonetheless, the reliability of these assessments is contingent upon the availability of effective and nutritionally adequate artificial diets. Our study demonstrates that even under 10-fold (7 μg/g) high-dose exposure, ds*AsFAR* did not negatively impact the development and survival of *H. axyridis*, a key predator within the cotton plant–pest–natural enemy triad. Similarly, dsRNA targeting *Brassicogethes aeneus* Fabricius (Coleoptera: Nitidulidae) (*dsalphaCOP*) exhibited no adverse effects on the parasitoid *Nasonia vitripennis* Walker (Hymenoptera: Pteromalidae) [34]. Additionally, no toxic effects were observed in other beneficial arthropods, such as *Telenomus podisi* Ashmead (Hymenoptera: Scelionidae) [35], *Propylea japonica* Thunberg (Coleoptera: Coccinellidae) [36,37], *Cyrtorhinus lividipennis* Reuter (Hemiptera: Miridae) [38,39], *Ontsira mellipes* Ashmead (Hymenoptera: Braconidae) [40], and *Coleomegilla maculata* DeGeer (Coleoptera: Coccinellidae) [41]. Nonetheless, certain studies indicate that insecticidal dsRNA may present species-specific risks to natural enemies. For instance, ds*vATPase-A*, which targets *Diabrotica virgifera* LeConte (Coleoptera: Chrysometidae), has been shown to significantly, albeit marginally, extend the developmental duration of the predator *Adalia bipunctata* Linnaeus (Coleoptera: Coccinellidae) and to significantly decrease the survival rate of another predator, *Coccinella septempunctata* Linnaeus (Coleoptera: Coccinellidae) [42,43]. These findings highlight the importance of conducting species-specific safety assessments within the plant–pest–natural enemy interaction framework to ensure a comprehensive ecological risk evaluation for RNAi-based crops. However, one limitation of our study is that we did not assess the multigenerational or long-term effects on *H. axyridis*, which merits further investigation.

Bioinformatics is a powerful method for identifying off-target effects. Typically, most contemporary research employs bioinformatics during the initial phase of selecting and designing target genes for transgenic dsRNA crops [19]. Specialized software, dsCheck [44] and OfftargetFinder [45], are utilized to conduct off-target searches and to design dsRNA with minimal sequence similarity to non-target genes, thereby reducing the possibility of unintended off-target effects [46]. However, due to the limited availability of sequence information for all species, it is not feasible to predict off-target effects solely through bioinformatics. As scientific research progresses, omics analysis techniques can be employed from a comprehensive perspective to evaluate the off-target effects of RNAi [47]. Additionally, RNAi induces specific effects on the transcriptomes of organisms, which may pose risks due to its potential adverse impacts, such as off-target gene silencing, on non-target species in RNAi-modified crops [21]. These off-target effects in numerous insects have been evaluated using omics technologies [21,32,36]. Taning et al. [48] conducted in silico predictions and identified numerous potential off-target genes when targeting dsRNA for the pollen beetle pest. However, when they tested these candidates in in vivo experiments, no negative effects on bumblebees were observed [48]. In this study, our transcriptomic analysis also identified unintended DEGs that may be induced by ds*AsFAR* in larvae and adults of *H. axyridis*. However, a limitation of this study is that to ensure consistency and minimize technical variation, the same RNA samples were used for both RNA-seq and qRT-PCR validation. This approach may introduce circularity and does not fully account for biological variability. Whether these unintended DEGs are truly induced by the inference of *AsFAR*, and whether further validation with independent RNA samples would yield consistent results, warrants further investigation in future ecological risk assessment. Additionally, our findings indicated that off-target effects were not correlated with the number of consecutive matched base pairs. RNAi could be triggered with as few as seven continuous matched base pairs between the target gene and exogenous dsRNA, as demonstrated in our previous study [21]. This implies that transcriptomes exhibit a high degree of adaptability to external dsRNA, necessitating the consideration of unintended transcriptomic alterations in the safety assessments of RNAi crops for non-target organisms. Nonetheless, these alterations alone do not comprehensively encapsulate potential risks [21]. The application of Shannon entropy could facilitate the evaluation of transcriptome stability [49] and may serve as one of the effective approaches for ecological risk assessment.

5. Conclusions

Compared to classical synthetic pesticides, RNAi offers an innovative, specific, and environmentally sustainable pest control method. Nonetheless, there are multiple challenges to address at the technical, regulatory, and risk assessment levels [20]. In this study, we proposed a biological and ecological risk evaluation framework for RNAi crops on insects' natural enemies, which includes assessing exposure routes for insecticidal dsRNA; testing food-chain delivery to reflect real-world conditions; conducting animal-feeding tests for worst-case effects; analyzing off-target effects for unintended silencing; and evaluating transcriptomic stability. In addition, our study also revealed that ds*AsFAR* did not exert any harmful impacts on *H. axyridis* at either the developmental or transcriptomic levels involving *FAR* or its related genes. In alignment with previous studies, our results highlight the importance of evaluating RNAi crops on an individual basis for their safety regarding non-target organisms, especially in relation to off-target consequences. Further investigation is necessary to understand the mechanisms involved. We hope this study will support the development of guidelines for biological and ecological risk assessment to minimize the potential negative consequences of RNAi crops under a plant–pest–natural enemy ecological interaction framework.

Supplementary Materials: The following supporting information can be downloaded at https://www.mdpi.com/article/10.3390/biology14091173/s1, Figures S1 and S2: The KEGG analysis of transcriptomes and the GO analysis of transcriptomes, respectively; Tables S1–S3: Primers used for this study, DEGs in *H. axyridis* adults fed ds*GFP* vs. ds*AsFAR*, and DEGs in *H. axyridis* larvae fed ds*GFP* vs. ds*AsFAR*, respectively.

Author Contributions: H.Y.: Conceptualization, investigation, formal analysis, visualization, writing—original draft; H.X.: Investigation; J.Y.: Conceptualization, formal analysis, writing—review and editing; W.M.: Conceptualization, writing—review and editing. All authors have read and agreed to the published version of the manuscript.

Funding: This research was funded by the Biological Breeding-Major Projects (No. 2023ZD04062) and the Start-up Funding from Wuhan Business University (No. ZXQD147).

Data Availability Statement: The raw sequence data are available in the China National Center for Bioinformation (https://ngdc.cncb.ac.cn/) under accession number PRJCA043540.

Acknowledgments: The computations in this paper were run on the bioinformatics computing platform of the National Key Laboratory of Crop Genetic Improvement, Huazhong Agricultural University.

Conflicts of Interest: The authors declare no conflicts of interest.

References

1. Itsathitphaisarn, O.; Thitamadee, S.; Weerachatyanukul, W.; Sritunyalucksana, K. Potential of RNAi applications to control viral diseases of farmed shrimp. *J. Invertebr. Pathol.* **2016**, *147*, 76–85. [CrossRef]
2. Zhao, J.H.; Guo, H.S. RNA silencing: From discovery and elucidation to application and perspectives. *J. Integr. Plant Biol.* **2022**, *64*, 476–498. [CrossRef]
3. Matzke, M.; Primig, M.; Trnovsky, J.; Matzke, A. Reversible methylation and inactivation of marker genes in sequentially transformed tobacco plants. *EMBO J.* **1989**, *8*, 643–649. [CrossRef]
4. Niu, D.; Hamby, R.; Sanchez, J.N.; Cai, Q.; Yan, Q.; Jin, H. RNAs—A new frontier in crop protection. *Curr. Opin. Biotech.* **2021**, *70*, 204–212. [CrossRef]
5. Stoutjesdijk, P.A.; Singh, S.P.; Liu, Q.; Hurlstone, C.J.; Waterhouse, P.A.; Green, A.G. hpRNA-mediated targeting of the Arabidopsis *FAD2* gene gives highly efficient and stable silencing. *Plant Physiol.* **2002**, *129*, 1723–1731. [CrossRef] [PubMed]

6. Luo, M.; Wang, Z.; Li, H.; Xia, K.F.; Cai, Y.; Xu, Z.F. Overexpression of a weed (*Solanum americanum*) proteinase inhibitor in transgenic tobacco results in increased glandular trichome density and enhanced resistance to *Helicoverpa armigera* and *Spodoptera litura*. *Int. J. Mol. Sci.* **2009**, *10*, 1896–1910. [CrossRef] [PubMed]
7. Heinemann, J.A. Should dsRNA treatments applied in outdoor environments be regulated? *Environ. Int.* **2019**, *132*, 104856. [CrossRef]
8. Heinemann, J.A.; Agapito-Tenfen, S.Z.; Carman, J.A. A comparative evaluation of the regulation of GM crops or products containing dsRNA and suggested improvements to risk assessments. *Environ. Int.* **2013**, *55*, 43–55. [CrossRef]
9. National Academies of Sciences; Medicine; Division on Earth; Committee on Genetically Engineered Crops; Past Experience; Future Prospects. *Genetically Engineered Crops: Experiences and Prospects*; National Academies Press: Washington, DC, USA, 2016.
10. Rengasamy, P. World salinization with emphasis on Australia. *J. Exp. Bot.* **2006**, *57*, 1017–1023. [CrossRef]
11. Wu, K.; Guo, Y. The evolution of cotton pest management practices in China. *Annu. Rev. Entomol.* **2005**, *50*, 31–52. [CrossRef] [PubMed]
12. Lu, Y.; Jiao, Z.; Wu, K. Early season host plants of *Apolygus lucorum* (Heteroptera: Miridae) in northern China. *J. Econ. Entomol.* **2012**, *105*, 1603–1611. [CrossRef]
13. Roberts, A.F.; Devos, Y.; Lemgo, G.N.; Zhou, X. Biosafety research for non-target organism risk assessment of RNAi-based GE plants. *Front. Plant Sci.* **2015**, *6*, 958. [CrossRef] [PubMed]
14. Luo, J.; Liang, S.; Li, J.; Xu, Z.; Li, L.; Zhu, B.; Li, Z.; Lei, C.; Lindsey, K.; Chen, L. A transgenic strategy for controlling plant bugs (*Adelphocoris suturalis*) through expression of double-stranded RNA homologous to fatty acyl-coenzyme A reductase in cotton. *New Phytol.* **2017**, *215*, 1173–1185. [CrossRef]
15. Fedorov, Y.; Anderson, E.M.; Birmingham, A.; Reynolds, A.; Karpilow, J.; Robinson, K.; Leake, D.; Marshall, W.S.; Khvorova, A. Off-target effects by siRNA can induce toxic phenotype. *Rna* **2006**, *12*, 1188–1196. [CrossRef]
16. Jackson, A.L.; Bartz, S.R.; Schelter, J.; Kobayashi, S.V.; Burchard, J.; Mao, M.; Li, B.; Cavet, G.; Linsley, P.S. Expression profiling reveals off-target gene regulation by RNAi. *Nat. Biotech.* **2003**, *21*, 635–637. [CrossRef]
17. Qiu, S.; Adema, C.M.; Lane, T. A computational study of off-target effects of RNA interference. *Nucleic Acids Res.* **2005**, *33*, 1834–1847. [CrossRef] [PubMed]
18. Shahrajabian, M.H.; Sun, W. Survey on multi-omics, and multi-omics data analysis, integration and application. *Curr. Pharm. Anal.* **2023**, *19*, 267–281. [CrossRef]
19. Zhou, Z.H.; Yao, Z.T.; Abouzaid, M.; Hull, J.J.; Ma, W.H.; Hua, H.X.; Lin, Y.J. Co-expression network analysis: A future approach for pest control target discovery. *J. Agric. Food Chem.* **2023**, *71*, 7201–7209. [CrossRef]
20. Martínez, O.; Reyes-Valdés, M.H. Defining diversity, specialization, and gene specificity in transcriptomes through information theory. *Proc. Natl. Acad. Sci. USA* **2008**, *105*, 9709–9714. [CrossRef]
21. Ma, W.; Wu, T.; Zhang, Z.; Hang, L.; Situ, G.; Yin, C.; Ye, X.; Chen, M.; Zhao, X.; He, K.; et al. Using transcriptome Shannon entropy to evaluate the off-target effects and safety of insecticidal siRNAs. *J. Integr. Agric.* **2022**, *21*, 170–177. [CrossRef]
22. Chen, Y.; De Schutter, K. Biosafety aspects of RNAi-based pests control. *Pest Manag. Sci.* **2024**, *80*, 3697–3706. [CrossRef] [PubMed]
23. Mamta, B.; Rajam, M. RNAi technology: A new platform for crop pest control. *Physiol. Mol. Biol. Plants* **2017**, *23*, 487–501. [CrossRef] [PubMed]
24. Thakur, N.; Upadhyay, S.K.; Verma, P.C.; Chandrashekar, K.; Tuli, R.; Singh, P.K. Enhanced whitefly resistance in transgenic tobacco plants expressing double stranded RNA of v-ATPase A gene. *PLoS ONE* **2014**, *9*, e87235. [CrossRef]
25. Jin, S.; Singh, N.D.; Li, L.; Zhang, X.; Daniell, H. Engineered chloroplast dsRNA silences *cytochrome p450 monooxygenase, V-ATPase and chitin synthase* genes in the insect gut and disrupts *Helicoverpa armigera* larval development and pupation. *Plant Biotechnol. J.* **2015**, *13*, 435–446. [CrossRef]
26. Tian, G.; Cheng, L.; Qi, X.; Ge, Z.; Niu, C.; Zhang, X.; Jin, S. Transgenic cotton plants expressing double-stranded RNAs target *HMG-CoA reductase* (*HMGR*) gene inhibits the growth, development and survival of cotton bollworms. *Int. J. Biol. Sci.* **2015**, *11*, 1296. [CrossRef] [PubMed]
27. Ni, M.; Ma, W.; Wang, X.; Gao, M.; Dai, Y.; Wei, X.; Zhang, L.; Peng, Y.; Chen, S.; Ding, L. Next-generation transgenic cotton: Pyramiding RNAi and Bt counters insect resistance. *Plant Biotechnol. J.* **2017**, *15*, 1204–1213. [CrossRef] [PubMed]
28. Hussain, T.; Aksoy, E.; Çalışkan, M.E.; Bakhsh, A. Transgenic potato lines expressing hairpin RNAi construct of molting-associated *EcR* gene exhibit enhanced resistance against Colorado potato beetle (*Leptinotarsa decemlineata, Say*). *Transgenic Res.* **2019**, *28*, 151–164. [CrossRef] [PubMed]
29. Mat Jalaluddin, N.S.; Othman, R.Y.; Harikrishna, J.A. Global trends in research and commercialization of exogenous and endogenous RNAi technologies for crops. *Crit. Rev. Biotechnol.* **2019**, *39*, 67–78. [CrossRef]
30. Christiaens, O.; Whyard, S.; Vélez, A.M.; Smagghe, G. Double-stranded RNA technology to control insect pests: Current status and challenges. *Front. Plant Sci.* **2020**, *11*, 451. [CrossRef]
31. Kleijn, D.; Bommarco, R.; Fijen, T.P.; Garibaldi, L.A.; Potts, S.G.; Van Der Putten, W.H. Ecological intensification: Bridging the gap between science and practice. *Trends Ecol. Evol.* **2019**, *34*, 154–166. [CrossRef]

32. Chen, J.; Wang, H.; Yang, X.; Chen, G.; Du, L.; Chen, H.; Li, Y.; Peng, Y.; Han, L. Consumption of miRNA-mediated insect-resistant transgenic rice pollen does not harm *Apis mellifera* adults. *J. Agric. Food Chem.* **2021**, *69*, 4234–4242. [CrossRef]
33. Kupferschmidt, K. A lethal dose of RNA. *Science* **2013**, *341*, 732733. [CrossRef]
34. Deveux, M.; Christiaens, O.; Smagghe, G.; Taning, N.T.C. *Nasonia vitripennis* as a parasitoid wasp model for pest-specific dsRNA safety assessment. *Entomol. Gen.* **2023**, *43*, 139–146.
35. Castellanos, N.L.; Smagghe, G.; Taning, C.N.T.; Oliveira, E.E.; Christiaens, O. Risk assessment of RNAi-based pesticides to non-target organisms: Evaluating the effects of sequence similarity in the parasitoid wasp *Telenomus podisi*. *Sci. Total Environ.* **2022**, *832*, 154746. [CrossRef]
36. Chen, J.; Sheng, C.-W.; Peng, Y.; Wang, K.; Jiao, Y.; Palli, S.R.; Cao, H. Transcript Level and Sequence Matching Are Key Determinants of Off-Target Effects in RNAi. *J. Agric. Food Chem.* **2023**, *72*, 577–589. [CrossRef]
37. Liu, Z.; Wang, Y.; Nanda, S.; Li, Z.; Li, Y.; Guo, M.; Chen, S.; Yang, C.; Pan, H. Oral delivery of ds*HvUSP* is a promising method for *Henosepilachna vigintioctopunctata* control with no adverse effect on the non-target insect *Propylea japonica*. *Entomol. Gen.* **2023**, *43*, 157–165. [CrossRef]
38. Dang, C.; Zhang, Y.; Sun, C.; Li, R.; Wang, F.; Fang, Q.; Yao, H.; Stanley, D.; Ye, G. dsRNAs Targeted to the Brown Planthopper *Nilaparvata lugens*: Assessing Risk to a Non-Target, Beneficial Predator, *Cyrtorhinus lividipennis*. *J. Agric. Food Chem.* **2022**, *70*, 373–380. [CrossRef] [PubMed]
39. Li, H.J.; Zhang, H.H.; Lu, J.B.; Zhang, C.X. Threonyl-tRNA synthetase gene, a potential target for RNAi-based control of three rice planthoppers. *Pest Manag. Sci.* **2022**, *78*, 4589–4598. [CrossRef]
40. Wang, X.; Faucher, J.; Dhandapani, R.K.; Duan, J.J.; Palli, S.R. Potential effects of RNA interference of Asian longhorned beetle on its parasitoid. *Pest Manag. Sci.* **2023**, *79*, 1557–1565. [CrossRef] [PubMed]
41. Hollowell, H.; Rieske, L.K. Southern pine beetle-specific RNA interference exhibits no effect on model nontarget insects. *J. Pest Sci.* **2022**, *95*, 1429–1441. [CrossRef]
42. Haller, S.; Widmer, F.; Siegfried, B.D.; Zhuo, X.; Romeis, J. Responses of two ladybird beetle species (Coleoptera: Coccinellidae) to dietary RNAi. *Pest Manag. Sci.* **2019**, *75*, 2652–2662. [CrossRef] [PubMed]
43. Pan, H.; Yang, X.; Romeis, J.; Siegfried, B.D.; Zhou, X. Dietary RNAi toxicity assay exhibits differential responses to ingested dsRNAs among lady beetles. *Pest Manag. Sci.* **2020**, *76*, 3606–3614. [CrossRef] [PubMed]
44. Naito, Y.; Yamada, T.; Matsumiya, T.; Ui-Tei, K.; Saigo, K.; Morishita, S. dsCheck: Highly sensitive off-target search software for double-stranded RNA-mediated RNA interference. *Nucleic Acids Res.* **2005**, *33* (Suppl. 2), W589–W591. [CrossRef]
45. Good, R.T.; Varghese, T.; Golz, J.F.; Russell, D.A.; Papanicolaou, A.; Edwards, O.; Robin, C. OfftargetFinder: A web tool for species-specific RNAi design. *Bioinformatics* **2016**, *32*, 1232–1234. [CrossRef]
46. Allen, M.L. Comparison of RNAi sequences in insect-resistant plants to expressed sequences of a beneficial lady beetle: A closer look at off-target considerations. *Insects* **2017**, *8*, 27. [CrossRef]
47. Kleter, G.A. Food safety assessment of crops engineered with RNA interference and other methods to modulate expression of endogenous and plant pest genes. *Pest Manag. Sci.* **2020**, *76*, 3333–3339. [CrossRef]
48. Taning, C.N.T.; Gui, S.; De Schutter, K.; Jahani, M.; Castellanos, N.L.; Christiaens, O.; Smagghe, G. A sequence complementarity-based approach for evaluating off-target transcript knockdown in *Bombus terrestris*, following ingestion of pest-specific dsRNA. *J. Pest Sci.* **2021**, *94*, 487–503. [CrossRef]
49. Carels, N. Assessing RNA-Seq workflow methodologies using Shannon entropy. *Biology* **2024**, *13*, 482. [CrossRef] [PubMed]

biology

MDPI

Article

A Ubiquitous Volatile in Noctuid Larval Frass Attracts a Parasitoid Species

Chaowei Wang [1,2,3,†], Xingzhou Liu [3,†], Sylvestre T. O. Kelehoun [1], Kai Dong [1], Yueying Wang [3], Maozhu Yin [3], Jinbu Li [3,4], Yu Gao [2,5,*] and Hao Xu [1,2,*]

1 State Key Laboratory of Agricultural and Forestry Biosecurity, College of Plant Protection, Nanjing Agricultural University, Nanjing 210095, China; 18815575879@163.com (C.W.); 2023102292@stu.njau.edu.cn (S.T.O.K.); 2023802253@stu.njau.edu.cn (K.D.)

2 Key Laboratory of Soybean Disease and Pest Control, Ministry of Agriculture and Rural Affairs, Nanjing Agricultural University, Nanjing 210095, China

3 Institute of Plant Protection, Suzhou Academy of Agricultural Sciences, Suzhou 234000, China; lxz8610@163.com (X.L.); wyysznky@163.com (Y.W.); 2018102025@njau.edu.cn (M.Y.); lijinbu0991@sina.com (J.L.)

4 Department of Agriculture, Suzhou Vocational and Technical College, Suzhou 234000, China

5 College of Plant Protection, Jilin Agricultural University, Changchun 130118, China

* Correspondence: gaothrips@jlau.edu.cn (Y.G.); haoxu@njau.edu.cn (H.X.)

† These authors contributed equally to this work.

Simple Summary: It is increasingly becoming common knowledge that predators and parasitoids use herbivore-induced plant volatiles (HIPVs) to locate their herbivore hosts. However, many natural enemies are also attracted to odors emitted directly from hosts or host-associated tissues. Little is known about the mechanism of the latter phenomenon. Here, we found that the larval bodies of two noctuid species had similar chemical compositions, as did their larval frass, which were also largely independent of the types of food (maize leaves versus an artificial diet without leaf tissues) supplied to the larvae. The volatile compound ethyl palmitate of frass was strongly attractive to the larval endoparasitoid *Microplitis mediator* (Haliday). However, the compound was rarely found in extracts of the larval bodies, which explains why the larval frass was more attractive to the parasitoids than their respective larval bodies. We also hypothesize that ethyl palmitate and some other compounds were common metabolites of gut digestion involving bacteria based on comparable analyses of samples obtained from the larvae fed diets supplemented with an antibiotic compound.

Abstract: Natural enemies commonly probe larval bodies and frass with their antennae for prey hunting. However, the attractants to natural enemies emitted directly from hosts and host-associated tissues remained largely unknown. Here, we used two generalist noctuid species, *Helicoverpa armigera* (Hübner) and *Spodoptera frugiperda* (JE Smith), along with the larval endoparasitoid *Microplitis mediator* (Haliday) to address the question. Extracts of larval frass of both the noctuid species were strongly attractive to *M. mediator* females when hosts were fed either maize, cotton, soybean leaves, or an artificial diet without leaf tissues. By using a combination of electrophysiological measurements and behavioral tests, we found that the attractiveness of frass mainly relied on a volatile compound ethyl palmitate. The compound was likely to be a by-product of host digestion involving gut bacteria because an antibiotic supplement in diets reduced the production of the compound in frass and led to the decreased attractiveness of frass to the parasitoids. In contrast, extracts of the larval bodies of both the noctuid species appeared to be less attractive to the parasitoids than their respective fecal extracts, independently of types of food supplied to the larvae. Altogether, larval frass of the two noctuid species was likely to be more important than their bodies in attracting the endoparasitoid species, and the main attractant of frass was probably one of the common metabolites of digestion involving gut microbes, and its emission is likely to be independent of host plant species.

Keywords: tritrophic interaction; parasitic wasp; foraging behavior; plant volatile; biological control; pest management

1. Introduction

Parasitoids are important agents in biological control programs against insect pests. They mainly rely on olfactory cues to locate their hosts in a complex chemical environment. Plant volatiles are released in large amounts induced by herbivory and serve as an important cue for parasitoids locating hosts [1]. However, herbivore-induced plant volatiles (HIPVs) are sometimes not host-specific (e.g., green leaf volatiles) [2–4], and host or host-associated cues are important for confirming the presence of hosts [5–7].

Odors of host by-products, such as frass, silks, and cocoons, often act as kairomones (benefiting natural enemies, but harmful to the hosts in this case) to attract parasitoids [8–14]. Since insect feces appear to be a reliable cue for many natural enemies to find prey/hosts, shelter-dwelling larvae of several lepidopteran families have evolved to perform an unusual behavior to avoid attacks by ejecting fecal pellets away from their feeding sites [15,16]. In addition, the smells of hosts are also attractive to parasitoids. For example, parasitoids often eavesdrop on host pheromones emitted by adults [5,17–19], and some larval parasitoids are directly attracted to host larval smells [20–24]. In the circumstances where HIPVs are not available, host and host-associated smells at the immature stage might be the principal cue for attracting parasitoid species, including those parasitizing pollinator hosts [9,25], belowground insects [20,26], and storage pests [27,28].

The attractiveness of host frass to parasitoids has often been found to be independent of the types of food consumed by polyphagous hosts, and both feces obtained from leaves and non-leaf contained artificial diets are attractive to parasitoid species [11,29,30]. Frass of hosts that are fed non-leaf plant tissues (e.g., roots, grains, and pollens) is also attractive to parasitoids [25–27]. These findings suggest that the attractants in feces are possibly common metabolites of host digestion, independently of types of host food. However, several studies have also revealed that ingested leaf tissues in host bodies and feces played an additional role in attracting natural enemies [19,31–34]. Even more, the attractiveness of frass is sometimes entirely dependent on the presence of the ingested leaf tissues, and frass obtained from non-leaf contained artificial diets is not attractive to parasitoids [31]. Therefore, identifying the attractants of frass is necessary for gaining a better understanding on how natural enemies use fecal cues for prey finding.

Recently, the roles of gut bacteria of hosts have received increasing attention in terms of mediating foraging behaviors of natural enemies. For example, gut bacteria of the Colorado potato beetle *Leptinotarsa decemlineata* Say produce defense compounds (e.g., mandelonitrile) to ward off natural enemies [35]. Gut bacteria are also responsible for producing some volatile compounds that mediate intraspecific aggregations in the German cockroach *Blattella germanica* (L.) [36]. When those microbe-related cues are reliable, it would be expected that parasitoid species might use them for locating hosts, as suggested in previous papers [37–40]. However, the specific compounds that are released because of activities of insect symbionts, but betray hosts, remain largely unknown.

The cotton bollworm *Helicoverpa armigera* (Hübner) (Lepidoptera: Noctuidae) and the fall armyworm *Spodoptera frugiperda* (JE Smith) (Lepidoptera: Noctuidae) are of great importance in agriculture. *H. armigera* occurs in many regions, including Europe, Africa, and Asia, and damages plant species of 68 families [41]. *S. frugiperda*, native to the New World, is now a devastating invasive pest of maize in Africa and Asia [42,43], and it also feeds on other crops, including rice, peanuts, and sorghum [44,45]. The solitary larval endoparasitoid *Microplitis mediator* (Haliday) (Hymenoptera: Braconidae) is widely distributed in Europe and Asia, and parasitizes a variety of noctuid species, including *H. armigera* and *S. frugiperda* [46–48]. This species is strongly attracted by HIPVs [46,49]. However, little is known about how the parasitoid species uses host/host-associated cues for locating hosts.

In this study, we first evaluated the relative attraction of the parasitoids to extracts of larval bodies and frass of the two noctuid species, taking into account the effects of food types supplied to the host larvae. By using a combination of chemical analyses and electroantennographic recordings, we identified a key compound that was responsible for the attractiveness to the parasitoids. Finally, we studied the relationship between the production of the compound and gut microbes by using an antibiotic supplement in diets.

2. Materials and Methods

Insects and Plants—The parasitoid species *M. mediator* first emerged from *H. armigera* caterpillars collected from soybeans grown in fields in Suzhou (33.636° N, 117.082° E), Anhui Province, China. When rearing, five female parasitoids were used to parasitize about 50 caterpillars of *H. armigera* (about 5 d old) that were kept in square plastic boxes (15 × 13 × 5 cm) and fed a wheat germ-based artificial diet (see below). Parasitism lasted for 24 h in a 25 °C incubator (LD 16:8 h), with honey provided as food for the female parasitoids, as described in a previous publication [50]. Wasp cocoons were collected from the plastic boxes about 2 weeks after parasitism and transferred to 30 × 30 × 30 cm nylon rearing tents (Hongrixing, Xiamen, China), and then provided with honey and moist cotton wool in a 25 °C incubator (LD 16:8 h). We used the naïve female wasps (2–7 d old after emergence) for olfactometer bioassays as described below.

The rearing of the host species *H. armigera* and *S. frugiperda* was initiated using caterpillars collected from soybean and maize plants, respectively, in Suzhou (33.636° N, 117.082° E), Anhui Province, China. Caterpillars of both these species were reared on a modified form of the wheat germ-based artificial diet as described in a previous paper [51]. Specifically, antimicrobial agents, including methyl paraben, sorbic acid, and chlortetracycline, were removed from the diet. Rearing was first carried out in the same plastic boxes used for parasitism, with 50 caterpillars placed in each box. The larvae were reared in groups before the fourth instar, and then individually in plastic tubes (3 × 5 cm) until pupation. The pupae of each species were transferred to the same type of rearing tent (30 × 30 × 30 cm) used to keep the wasps, and we provided emerged moths with 10% honey solution as food.

Seedlings of maize, cotton, and soybean were planted in plastic tubes (OD = 4.5 cm, L = 13 cm) and grown for three weeks in an incubator (28 °C, LD = 16:8 h). The leaves were harvested daily to feed caterpillars of both the species, starting with 50 individuals (4-d old) in the square plastic boxes. Thus, caterpillars of the both species were fed either maize, cotton, soybean leaves, or the artificial diet. After 6 days of feeding, frass and caterpillars were collected and used for odor extraction (see below).

Sample collections on diets supplemented with the antibiotic—Two micrograms of tetracycline (98%, Macklin Ltd., Shanghai, China) were added to one liter of the artificial diet when the temperature of the liquid diet had cooled to 55 °C during diet preparation and the amount of the antibiotic compound was similar to that of other antimicrobial agents of the artificial diet described by Merkx-Jacques and Bede [51]. In addition, 20 µg of tetracycline that was dissolved in 1 mL of water was evenly sprayed on leaves of maize seedlings (2–3 weeks old), and the control maize seedlings were applied 1 mL of water, similar to a previous paper [37]. Fifty caterpillars (about 5 d old after hatching) were weighed, and then we fed them with one of the two types of food (maize leaves or the artificial diet) supplemented either with or without the antibiotic compound in the square plastic boxes (15 × 13 × 5 cm). The frass and caterpillars were collected twice from the boxes after 3 d or 6 d of feeding for odor extraction (see below). The caterpillars were weighed again in groups of ten individuals, and their weight gain rates were calculated. The data were analyzed with the *t*-test performed by SigmaPlot 12.5.

Solvent extraction—After the two periods (the first or the second 3 d periods, i.e., 1st–3rd and 4th–6th d) of feeding, the larvae that were randomly chosen from the rearing boxes were killed by placing them in a freezer at −20 °C for 10 min. Compounds in 100 mg of caterpillars (about 10 larvae after 6 d of feeding and 20 larvae after 3 d of feeding) and in 100 mg of their frass were extracted, respectively, by adding 500 μL of dichloromethane (DCM) (n = 5). During extraction, the samples were vibrated for about 30 s on a Vortex-Genie 2 mixer (Scientific Industries, Bohemia, NY, USA). The liquid supernatant was removed with a syringe and stored in vials at −20 °C before performing bioassays or chemical analyses.

Bioassays—A four-arm olfactometer was used to test the attractiveness of frass and caterpillar extracts to the parasitic wasps as previously described [52]. Each olfactometer arm delivered a gentle stream of air (0.6 L/min) into the central release chamber. The naïve female wasps were released in groups of six into the olfactometer, where they were allowed to choose among the treatments for up to 30 min. At the end of this period, or as soon as all six wasps had made a choice, the result was recorded. Then, the tested wasps were removed, and a new group of six naïve wasps was released into the olfactometer. Each replicate included four releases of six wasps. Each experiment consisted of six replicates (144 wasps in total, n = 6). Glassware was cleaned using the method described by Desurmont et al. [53], and the positions of the treatments in the olfactometer were changed randomly for each replicate.

Samples (10 μL) were applied on paper strips (1 × 3 cm) that had been placed on two opposite arms of the four-arm olfactometer, and the rest arms in between were treated with the same volume of control solvent (DCM) likewise. The design aimed to minimize the interactions between the tested parasitoids in the olfactometers [54]. Before testing, the paper strips were placed into a running fume hood for 15 min to allow the solvent to evaporate [55].

The data obtained from the olfactometer test did not meet the normality assumptions. Therefore, a Poisson generalized linear mixed model (GLMM) was run to assess the effects of the treatments (fixed factor) on the insect choices, with replicates of the experiment as a random blocking factor [56], which was performed in R version 3.0.2 with the package lme4 [57].

Gas Chromatography–Mass Spectrometry Analyses—The samples were analyzed with an Agilent 8890 gas chromatograph (GC), coupled with an Agilent 5977B mass spectrometer (MS) detector (Agilent Technologies Inc., Palo Alto, CA, USA). For each sample, 2 μL of extract was injected into the GC column using 10% split mode. The temperature of injector was 250 °C. Helium was used as the carrier gas within a non-polar column (HP-5ms, 30 m, 0.25 mm ID, 0.25 μm film thickness, Agilent Technologies) at a constant flow rate (1 mL/min). The oven temperature was initially set to 40 °C, held for 7 min, and then increased at a rate of 5 °C per minute to 55 °C, followed by a ramp of 15 °C per minute to 250 °C, and then held at this temperature for 2 min. The compounds were first identified by mass spectrometry analysis by comparing their mass spectra with the NIST mass spectral library database (Version 2.4). Most compounds were then confirmed by injection of commercially available standards, including (2*S*,3*S*)-(+)-butanediol (98%, Bidepharm Ltd., Shanghai, China), meso (*R*,*S*)-2,3-butanediol (99%, Macklin Ltd., Shanghai, China), pentadecane (99%, Macklin, Shanghai), palmitoleic acid (98%, Macklin, Shanghai), palmitic acid (98%, Mreda, Beijing, China), benzaldehyde (98%, Macklin, Shanghai), and ethyl palmitate (98%, Macklin, Shanghai). However, the identifications of 2-ethylhexanol and (*E*)-coniferol were considered as tentative based on suggestions by the NIST database.

Two internal standards n-octane and nonyl acetate (200 ng each combined in 10 μL dichloromethane) were added to each sample as references and were used to quantify the collected compounds. Pairwise comparisons of the compound amounts of extracts were carried out with the *t*-test performed using SigmaPlot 12.5.

Gas Chromatography–Electroantennographic Detection (GC-EAD) Analyses—The GC-EAD system consisted of an Agilent 6890 GC, equipped with a flame ionization detector (FID) and a non-polar column (HP-5ms, 30 m, 0.25 mm ID, 0.25 μm film thickness, Agilent Technologies), and was coupled to an EAD setup (Syntech, Hilversum, The Netherlands). Aliquots (2 μL) of the samples were injected splitless into the GC injector (280 °C) at an initial column temperature of 40 °C, and then the temperature was increased at a rate of 15 °C per minute to 250 °C. The GC effluent was split (split ratio FID:EAD = 1:3) by using a μFlow splitter (Gerstel, Mühlheim, Germany), and 25 mL/min of make-up gas (nitrogen) was added to carry the separated compounds through two deactivated capillaries: one leading to the FID, and the other leading to the EAD setup. The EAD outlet entered into purified and humidified constant airflow that was directed over the antennal preparation.

To prepare the antennae, *M. mediator* females were anesthetized with CO_2, and their heads were pulled out from the body with forceps, and the two tips of both antennae were cut and inserted into a glass capillary [58]. To close the electric circuit, the severed head was connected via the neck to another glass capillary. The capillaries were filled with insect Ringer solution (5 g NaCl; 0.42 g KCl; 0.19 g $CaCl_2$ in 1000 mL demineralized water). Each EAD test was replicated five times with different insect heads.

3. Results

The *M. mediator* females were more strongly attracted to the larval frass of both the noctuid hosts compared to the solvent controls when the caterpillars were fed either maize, cotton, soybean leaves (GLMM, $p < 0.001$, $n = 6$), or the artificial diet ($p < 0.01$) (Figure 1a,b). In both the species, the frass obtained from the larvae fed maize leaves was more attractive to the parasitoids than the frass obtained from the larvae fed the artificial diet (GLMM, $p < 0.05$, $n = 6$) (Figure 1a,b). Similarly, frass obtained from the *H. armigera* larvae fed the cotton leaves was more attractive to the parasitoids than that obtained from larvae fed the artificial diet (Figure 1a). However, the frass derived from soybean leaves was as attractive as that derived from the artificial diet, regardless of the species (GLMM, $p > 0.05$) (Figure 1a,b).

Similar tests on extracts of the larval bodies indicated that the body odors of both the host species were not significantly attractive to the parasitoids when the larvae were fed either cotton, soybean leaves, or the artificial diet (GLMM, $p > 0.05$, $n = 6$) (Figure 1c,d). However, when the larvae of both the species were fed maize leaves, the attractiveness of their body extracts to the parasitoids was significantly increased as opposed to the solvent control (GLMM, $p < 0.05$) (Figure 1c,d).

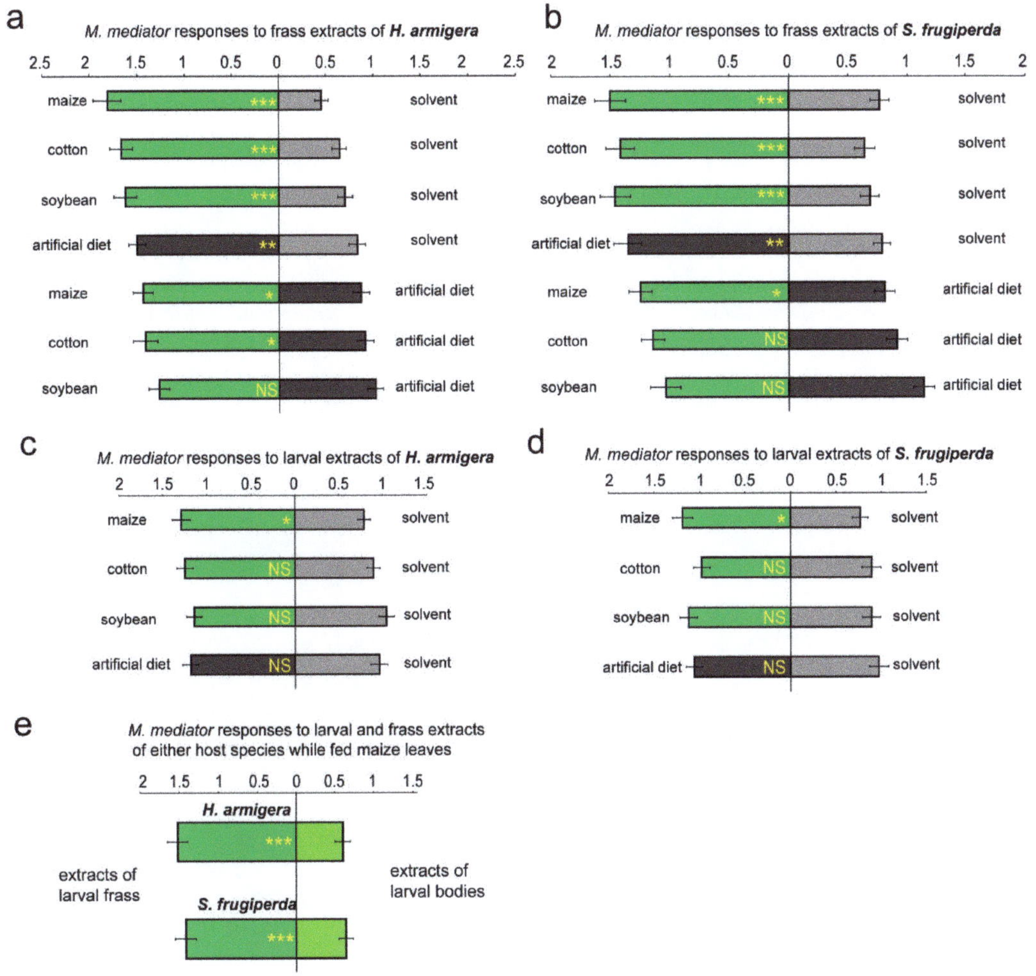

Figure 1. Attractiveness of solvent extracts of the larval bodies and frass of the two noctuid species to the parasitoid *M. mediator*. The response of *M. mediator* to odors of the larval frass of *H. armigera* (**a**) or *S. frugiperda* (**b**) when their larvae were fed either maize, cotton, soybean leaves, or an artificial diet without leaf tissues. The same volume (10 μL) of dichloromethane was used as the solvent control. Similarly, the response of *M. mediator* to odors of the larval bodies of *H. armigera* (**c**) or *S. frugiperda* (**d**) was also tested when the larvae were fed various types of food. (**e**) The relative attractiveness to *M. mediator* of the larval frass and bodies was compared when the larvae of either species were fed maize leaves. The asterisks on the bar graphs indicate significant differences between the treatments (GLMM, $n = 6$; *: p as 0.01~0.05, **: p as 0.001~0.01, ***: $p < 0.001$). "NS" indicates no significant difference between the treatments.

To obtain a clearer vision of the relative importance of the larval frass and bodies in attraction of the parasitoids, we also compared the attractiveness of the larval bodies and frass to the parasitoids when the larvae were fed maize leaves, and found that the attractiveness of frass was significantly stronger, independently of the host species (GLMM, $p < 0.001$, $n = 6$) (Figure 1e). The comparable tests were not performed on the samples derived from the other types of food (including soybean, cotton leaves, and the artificial diet) because the attractiveness of the larval bodies given these diets was not significantly stronger than that of the solvent control (GLMM, $p > 0.05$, $n = 6$) (Figure 1c,d).

The chemical compositions of larvae and frass were different. For example, the alkane compound pentadecane was the main compound of caterpillars of both the noctuid species, but it was released in much lower amounts in fecal extracts of both the species (*t*-test, $p < 0.05$, $n = 5$) (Figure 2 and Table 1). In contrast, two 2,3-butanediol stereoisomers, two fatty acids (palmitoleic acid and palmitic acid), and the ester compound ethyl palmitate were emitted in higher amounts by the frass of both the species than by the larval bodies (*t*-test, $p < 0.05$) (Figure 2 and Table 1). The diet types also led to different amounts of compounds produced by the frass of both the species. Three compounds, i.e., benzaldehyde, 2-ethylhexanol, and (*E*)-coniferol, were exclusively found in the frass obtained from the larvae fed maize leaves, but not in the frass of those fed the artificial diet (Figure 2 and Table 1; they are referred to as leaf-associated volatiles in this paper). However, amounts of the other compounds of the larval frass and bodies were independent of the diet types in both the noctuid species (Table 1, *t*-test, $p > 0.05$, $n = 5$).

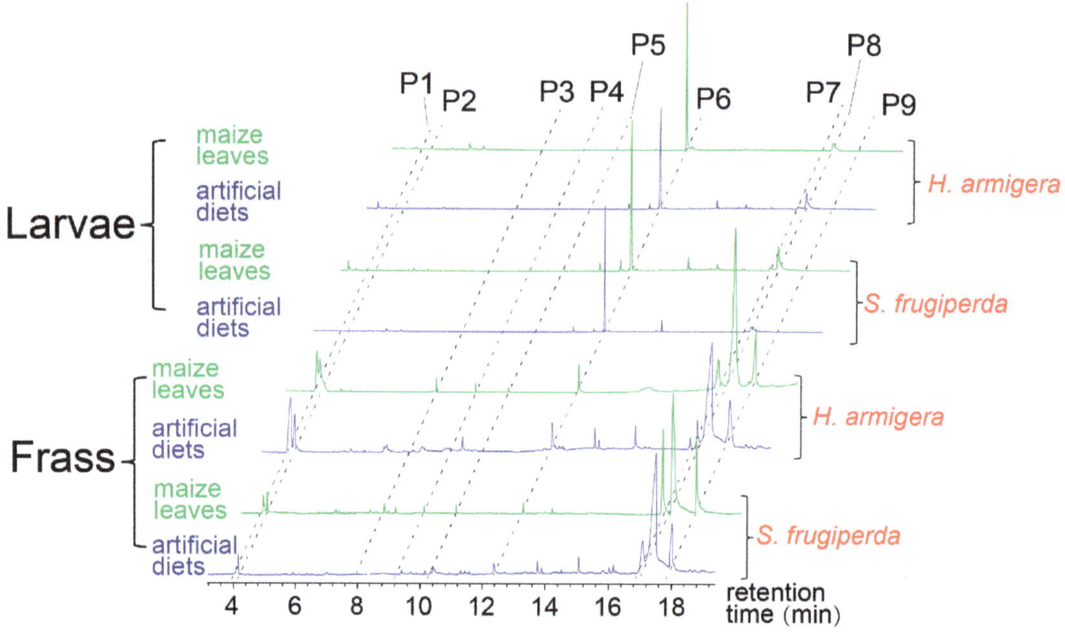

Figure 2. Typical chromatographs of the solvent extracts of larval bodies or frass of both the noctuid species when the larvae were fed either maize leaves or the non-leaf contained artificial diet. P1 and P2: two stereoisomers of 2, 3-butanediol; P3: benzaldehyde; P4: 2-ethylhexanol; P5: (*E*)-coniferol; P6: pentadecane; P7: palmitoleic acid; P8: palmitic acid; P9: ethyl palmitate. The amounts of these compounds are presented in Table 1. The two 2,3-butanediol stereoisomers are typical VOCs of bacteria [59], and benzaldehyde, 2-ethylhexanol, and (*E*)-coniferol are considered as leaf-associated volatiles in this paper because they were only found in the frass obtained from the larvae fed maize leaves, but not in frass of those fed the artificial diet.

The GC-EAG analyses showed that ethyl palmitate in the fecal extracts of both the species stimulated a strong antennal response in the parasitoids (Figure 3c,d), whereas the extracts of larval bodies did not elicit apparent EAG responses (Figure 3a,b). The commercial standard of ethyl palmitate elicited a similar antennal response (Figure 3e). The behavioral tests indicated that the compound was strongly attractive to the parasitoids

(Figure 3f; GLMM, *: p as 0.01~0.05, ***: $p < 0.001$, $n = 6$), with a positive change in relation to the amounts of compound applied (GLMM with Tukey's post-hoc test, dose effect: $F_{1,188} = 8.2$, $p = 0.005$).

The frass collected from the caterpillars fed maize leaves supplemented with tetracycline was significantly less attractive to the parasitoids after both the early (first–third days) and the late (fourth–sixth days) feeding periods compared to the frass from the larvae fed control maize leaves (Figure 4a,b; GLMM, *: p as 0.01~0.05, **: p as 0.001~0.01, ***: $p < 0.001$, $n = 6$). Similar results were also obtained when tetracycline was added to the artificial diet, although the reduction in attractiveness became significant only after 6 days of feeding in both the species (Figure 4a,b), suggesting that the longer feeding duration (6 d) on the tetracycline-treated diets led to a stronger reduction in attractiveness to the parasitoids (Figure 4a,b; GLMM with Tukey's post-hoc test, effects of feeding days: $F_{1,188} = 4.1$~7.4, $p < 0.05$; however, no such difference appeared in *H. armigera* when the larvae were fed maize leaves: $F_{1,188} = 1.5$, $p = 0.216$).

Table 1. Amounts (ng) of compounds in larval frass or bodies of the two noctuid species when they were fed either maize leaves or the artificial diet for 6 days.

Compounds	*Ha* Frass		*Sf* Frass		*Ha* Larvae		*Sf* Larvae	
	Maize	AD	Maize	AD	Maize	AD	Maize	AD
2,3-butanediol I ▲	63 ± 16 a	95 ± 13 a	23 ± 7 a′	11 ± 4 a′	n.d.	n.d.	n.d.	n.d.
2,3-butanediol II ▲	42 ± 12 a	54 ± 17 a	27 ± 9 a′	21 ± 8 a′	n.d.	n.d.	n.d.	n.d.
Benzaldehyde ◇	13 ± 2	n.d.	11 ± 3	n.d.	n.d.	n.d.	n.d.	n.d.
2-ethylhexanol ◇	10 ± 2	n.d.	14 ± 5	n.d.	n.d.	n.d.	n.d.	n.d.
(*E*)-coniferol ◇	7 ± 1	n.d.	16 ± 5	n.d.	n.d.	n.d.	n.d.	n.d.
Pentadecane B	17 ± 6 a	19 ± 7 a	16 ± 4 a′	13 ± 3 a′	**183 ± 21 a″**	**156 ± 28 a″**	**245 ± 62 a‴**	**215 ± 31 a‴**
palmitoleic acid	47 ± 10 a	41 ± 13 a	24 ± 10 a′	26 ± 6 a′	n.d.	n.d.	n.d.	n.d.
palmitic acid F	**271 ± 51 a**	**254 ± 34 a**	**286 ± 45 a′**	**213 ± 31 a′**	21 ± 6 a″	33 ± 9 a″	18 ± 6 a‴	9 ± 3 a‴
ethyl palmitate	81 ± 12 a	73 ± 11 a	93 ±17 a′	74 ± 22 a′	n.d.	n.d.	n.d.	n.d.

Pairwise comparisons were conducted with the *t*-test to analyze the differences in the amounts of compounds of the larval frass when larvae of *Helicoverpa armigera* (abbr. as *Ha*) and *Spodoptera frugiperda* (abbr. as *Sf*) were fed either maize leaves or the artificial diet. Similar analyses were likewise performed on the data of the larval bodies. As a result, the amounts of two 2,3-butanediol stereoisomers, pentadecane, palmitoleic acid, palmitic acid, and ethyl palmitate were independent of types of food in the frass or body extracts of both the species ($p > 0.05$, $n = 5$, indicated by the parallel symbols of a, a′, a″, and a‴ in cells). However, three compounds (indicated by "◇" in cells) were found in the frass of both the species when their larvae were fed maize leaves, but were not detected (abbr. as n.d. in cells) in the frass of those fed the artificial diet or in the body samples. The superscript letter "B" indicates that the larval bodies released significant higher amounts (bold numbers) of the compounds than did their respective larval frass, regardless of the diet types. On the contrary, "F" indicates that the larval frass included higher amounts (bold numbers) of compounds than did the corresponding bodies (i.e., pairwise comparisons between the body and frass samples of either species fed one of the two diets, *t*-test, $p < 0.05$, $n = 5$). (▲) 2,3-butanediol I was identified as (2*S*,3*S*)-(+)-butanediol, and 2,3-butanediol II was meso (*R*,*S*)-2,3-butanediol.

Additions of tetracycline to maize leaves or the artificial diet ceased the production of the two 2,3-butanediol stereoisomers in frass of both the species after 6 days of feeding (Figure 4c–f). Meanwhile, the production of ethyl palmitate was also reduced (*t*-test, $p < 0.05$, $n = 5$). However, the amounts of the three leaf-associated volatiles (i.e., benzaldehyde, 2-ethylhexanol, and (*E*)-coniferol), pentadecane, and the two fatty acids were not significantly affected by dietary addition of tetracycline in both types of food (*t*-test, $p > 0.05$, $n = 5$).

Providing the artificial diet or maize leaves that had been supplemented with tetracycline led to lower weight gain rates of caterpillars of both the species, and the differences were generally statistically significant after 6 d of feeding (*t*-test, $p < 0.001$, $n = 5$) (Figure 5).

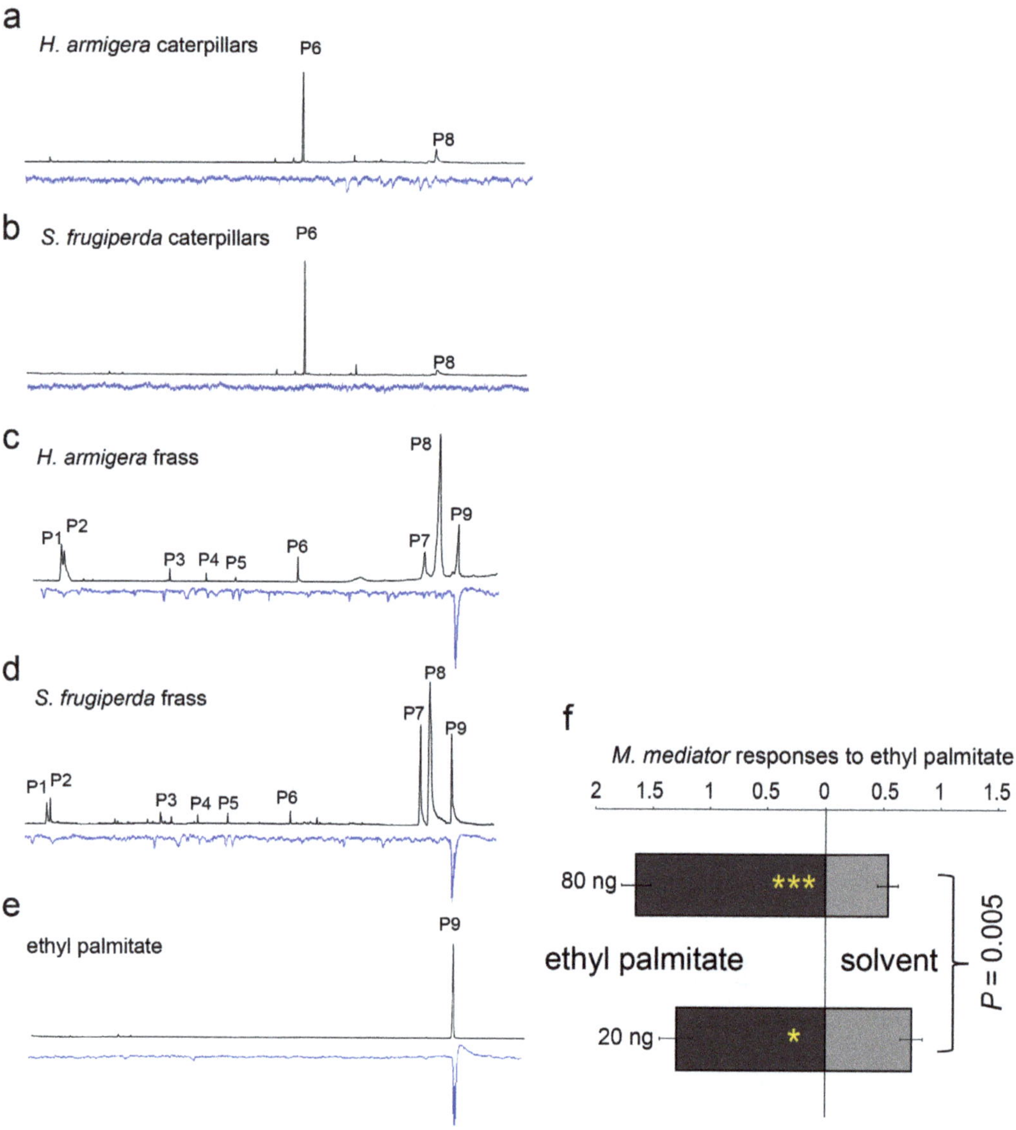

Figure 3. Electrophysiological responses of the antennae of the *M. mediator* females to solvent extracts of the *H. armigera* or *S. frugiperda* caterpillars (**a**,**b**) or frass (**c**,**d**) when their larvae were fed maize leaves. In addition, the electrophysiological response to the synthetic standard ethyl palmitate was also tested (**e**). Attractiveness of ethyl palmitate to the parasitoid *M. mediator* females at various doses was tested, where the same volume (10 μL) of dichloromethane was used as the solvent control (**f**) (GLMM with Tukey's post-hoc test, $n = 6$; *: p as 0.01~0.05, ***: $p < 0.001$).

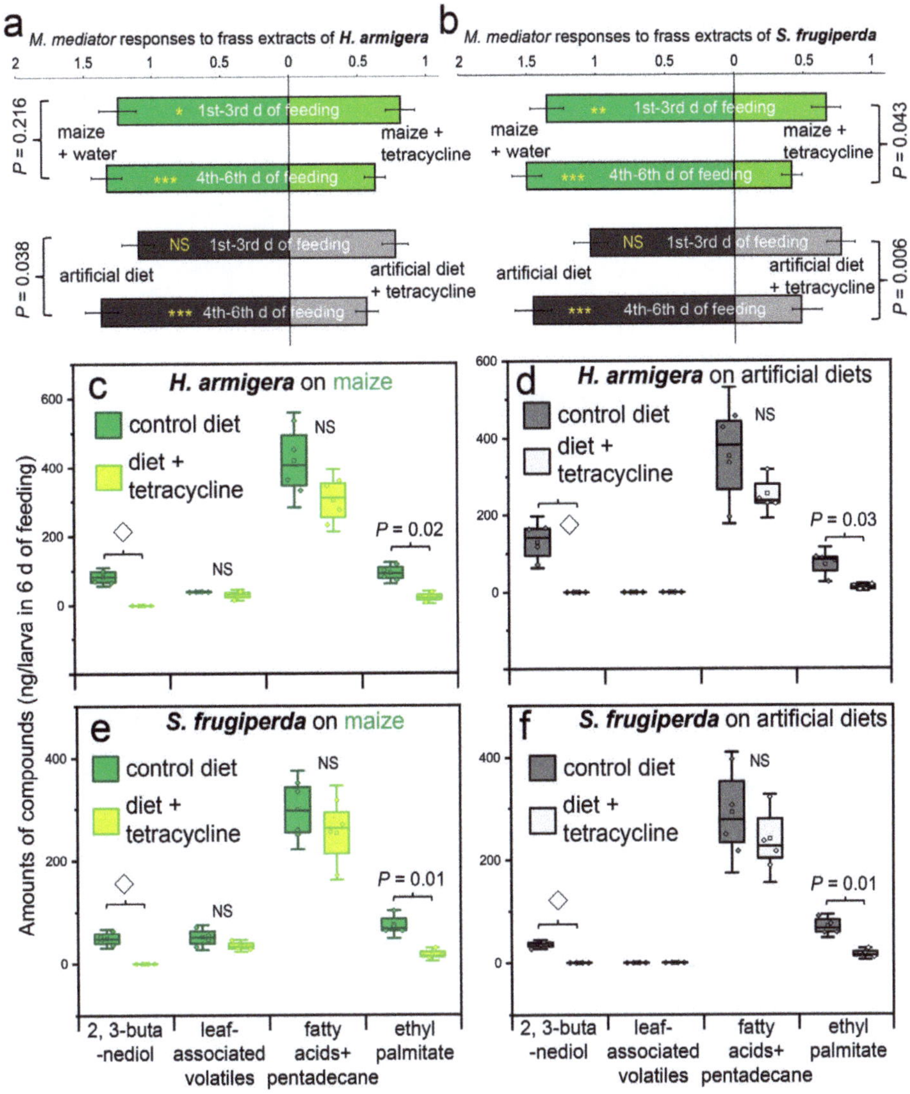

Figure 4. Attractiveness of the solvent extracts of larval frass of the two noctuid species (**a**,**b**) to the parasitoid *M. mediator* when the larvae were fed maize leaves or the artificial diet supplemented with the antibiotic compound tetracycline. The feeding of larvae lasted either 3 or 6 days (i.e., 1st~3rd or 4th~6th days of feeding). The asterisks on the bar graphs indicate significant differences between the treatments (GLMM with Tukey's post-hoc test, $n = 6$; *: p as 0.01~0.05, **: p as 0.001~0.01, ***: $p < 0.001$). "NS" indicates no significant difference between the treatments. The amounts of compounds found in larval frass of the two noctuid species (**c**,**d** for *H. armigera*; **e**,**f** for *S. frugiperda*) when they were fed one of two types of food (the maize leaves shown in **c**,**e**; the artificial diet shown in **d**,**f**) that had been (or not been) supplemented with tetracycline for 6 days. The differences were analyzed with the *t*-test ($n = 4$). "◇" is used to highlight that the compounds were not detected in the frass obtained from the larvae fed diets supplemented with tetracycline.

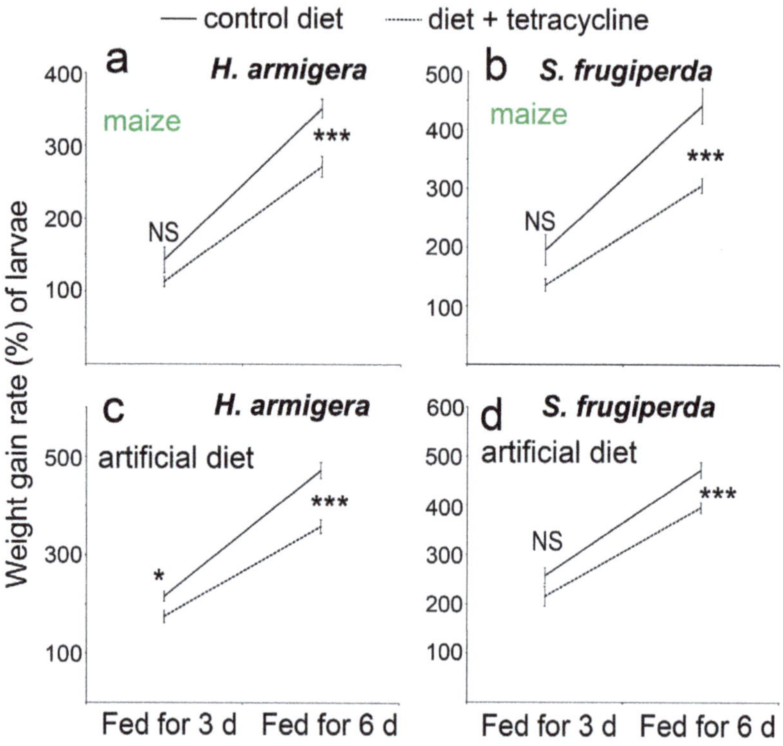

Figure 5. The weight gain rates (%) of caterpillars of the two noctuid species when the two types of food (maize leaves shown in **a**,**b** and the artificial diet shown in **c**,**d**) were supplemented with the antibiotic compound tetracycline. Feeding lasted 3 or 6 days. The differences were analyzed with the *t*-test ($n = 5$, *: p as 0.01~0.05, ***: $p < 0.001$). "NS" indicates no significant difference between the treatments.

4. Discussion

Parasitoid species rely on various odor cues to locate their host species, which include HIPVs, host body odors, and host-associated smells, such as those emitted from silks and frass [1,9]. Host fecal odors have often been reported as a reliable cue for natural enemies to locate their hosts or prey [10–16]. Unlike eavesdropping on host pheromones in a relatively long range [5,18], natural enemies normally sense host fecal odors in a relatively short range. For example, using antennae to probe frass (compounds) is a common behavior of parasitoid species in confirming the presence of host species [11,60]. Although such interactions between hosts and parasitoids are common, the underlying mechanisms remain largely unexplored. By combining behavioral bioassays, chemical analyses, and electrophysiological analyses, we found that the ester compound ethyl palmitate was responsible for the attractiveness of larval frass of the two noctuid species to the parasitoid *M. mediator*.

Relative importance of smells of host frass and bodies in attracting parasitoids—Many studies have found that odors of host bodies and/or frass are attractive to parasitoid species [20,21,34,61]. However, the relative importance of these two kinds of odor in attracting parasitoids has rarely been compared and discussed. Shelter-dwelling larvae of several lepidopteran families have evolved to perform an unusual behavior to avoid attacks from natural enemies by ejecting fecal pellets away from their homes [15,16], suggesting that larval frass of some lepidopterans is important in attracting natural enemies, as also confirmed in the present study. We also found that the odors of larval bodies were not significantly attractive to the parasitoids in general, although the presence of leaf tissues (e.g., maize) in larval bodies sometimes increased attractiveness to the parasitoids. Therefore, this study concluded that larval frass of the two noctuid species was more important than host bodies in attracting the parasitoids. This conclusion is in line with some previous studies that also found frass to be more attractive to natural enemies than host bodies [20,26,30].

However, some parasitoid species are not significantly attracted to their host's frass [22,62], and some other signals, such as visual cues and vibrations, possibly work in confirming the presence of host species by parasitoids [62,63]. Thus, the success of host foraging may depend on various cues, with variations according to parasitoid species [64].

The underlying mechanism of why fecal attractiveness to parasitoids does not depend on food types—The frass obtained from larvae fed various types of food, including maize, cotton, soybean leaves, and the artificial diet, was strongly attractive to the parasitoid *M. mediator*, suggesting that the attractiveness was largely independent of the larval diets. Similar results have also been observed in interactions between other generalist herbivores and their parasitoids [11,30]. However, the presence of ingested leaf tissues (e.g., maize) in larval bodies and frass increased the level of attractiveness to *M. mediator*, highlighting that leaf residues of some plant species in larval bodies and frass indeed play an additional role in attraction of natural enemies, as previously suggested [19,31,32].

The innate attractiveness of larval frass to parasitoids even occurs in non-host species. For example, the braconid parasitoid *Cotesia rubecula* is attracted by frass of the non-host species *Pieris napi*, and the level of attractiveness was as strong as that of the host species *Pieris rapae* [10]. These results indicate that the attractiveness of frass to *C. rubecula* relied on some conserved compounds that remained unchanged in the phylogenetically related species.

The chemical compositions of larval frass that was obtained from the insects fed one of the two types of diet could reveal the underlying mechanism. Although leaf-associated volatiles (i.e., benzaldehyde, 2-ethylhexanol, and (*E*)-coniferol) were only found in the frass obtained from the larvae fed maize leaves, but not in the frass from those given the artificial diet, the other compounds were released in similar amounts by the frass from both the species fed either type of food (maize leaves versus the artificial diet) (Table 1). Among those compounds, ethyl palmitate elicited a strong EAG response in *M. mediator* and attracted the parasitoids.

Palmitic acid and ethyl palmitate are possibly common by-products of insect digestion—Palmitic acid is one of the most abundant saturated fatty acids in many biological organisms, including plants [65]. The compound or its derivatives (e.g., ethyl palmitate) are sometimes found in frass (or other anal secretions) of herbivorous insects, including stink bugs, fruit flies, and beetles [66–68]. Our data showed that palmitic acid and ethyl palmitate also occurred in larval frass of the noctuid species. As mentioned above and shown in Table 1, the proportions (or amounts) of these compounds remained relatively similar across the frass from the larvae fed the two types of food, possibly because they were by-products of larval digestion of plant tissues.

In addition, gut bacteria that have been believed to be essential in digestion in insects [69] also contributed to greater weight gains in the two noctuid larvae, which was apparent after a relatively long period of feeding (6 days), but generally not statistically significant after a relatively short period of feeding (3 days). This was possibly because the gut bacteria completely lost their function only after several days of the antibiotic treatment. However, we observed that the survival rates were not reduced after 6 days of feeding on the tetracycline-treated diet.

How might VOCs of gut symbionts mediate predator–prey interactions?—Microbes are talented chemists in producing large amounts of VOCs with high compound diversity, diverse metabolic pathways and multiple ecological roles [70]. When they become insect symbionts, their VOCs are expected to imbue host insects with exceptional smells that have multiple functions in interactions with their natural enemies. For example, mandelonitriles released by gut bacteria of the beetle *L. decemlineata* are used by the insect hosts to defense against natural enemies [35]. In contrast, VOCs of gut bacteria often betray hosts by attracting natural enemies [39,40,71], as also found in this study. Two stereoisomers of 2,3-butanediol are the typical VOCs of bacteria [59]. In our study, we found that they also occurred in the larval frass of both the noctuid species. Interestingly, supplementing their diets with tetracycline completely ceased the release of the compounds from the larvae frass in both the moth species. Meanwhile, the amounts of the attractant ethyl palmitate were also reduced. These findings introduce the idea that gut digestion, involving bacteria, contributes to the formation of VOCs of frass, which shape important predator–prey interactions, as found here and in several other studies [35,37].

5. Conclusions

The chemical compositions of the noctuid larval bodies and frass remained relatively similar, which were also largely independent of the diet types. Larval frass, rather than the bodies of both the species, was significantly attractive to the larval endoparasitoid *M. mediator.* A ubiquitous compound, ethyl palmitate, found in the frass of both the species was responsible for the attraction of *M. mediator* and was likely to be one of common metabolites of gut digestion involving symbionts. This study contributes to a better understanding of host–parasitoid interactions, which might be important for the developments of biological control programs in the future, for example, by using ethyl palmitate as an attractant to natural enemies for enhancing biological control services on crops.

Author Contributions: H.X., X.L., J.L. and Y.G. designed the experiments. C.W., S.T.O.K., K.D., X.L., Y.W. and M.Y. performed the bioassays; S.T.O.K. and H.X. carried out the chemical analyses. H.X. analyzed data and made the figures. H.X. wrote an early version of this manuscript. All authors commented on this manuscript. All authors have read and agreed to the published version of the manuscript.

Funding: This research was funded by the Natural Science Foundation of Jiangsu Province, China (BK20221001 for Hao Xu); the Natural Science Fund of Education Department of Anhui Province, China (2023AH040365 for Jinbu Li); and the Earmarked Fund for China Agriculture Research System of MOF and MARA (Grant No. CARS04 for Yu Gao).

Institutional Review Board Statement: Not applicable.

Informed Consent Statement: Not applicable.

Data Availability Statement: All data used in this manuscript were archived on Mendeley Data at https://data.mendeley.com/drafts/wyrsb3py54 (accessed on 6 July 2025) (DOI: 10.17632/wyrsb3py54.1).

Acknowledgments: Ying Zheng from our institute, Nanjing Agricultural University, helped us with GC analyses.

Conflicts of Interest: The authors declare no competing interests.

References

1. Turlings, T.C.; Erb, M. Tritrophic interactions mediated by herbivore-induced plant volatiles: Mechanisms, ecological relevance, and application potential. *Annu. Rev. Entomol.* **2018**, *63*, 433–452. [CrossRef]
2. Turlings, T.C.J.; McCall, P.J.; Alborn, H.T.; Tumlinson, J.H. An elicitor in caterpillar oral secretions that induces corn seedlings to emit chemical signals attractive to parasitic wasps. *J. Chem. Ecol.* **1993**, *19*, 411–425. [CrossRef] [PubMed]

3. Reinecke, A.; Ruther, J.; Tolasch, T.; Francke, W.; Hilker, M. Alcoholism in cockchafers: Orientation of male *Melolontha melolontha* towards green leaf alcohols. *Naturwissenschaften* **2002**, *89*, 265–269. [CrossRef] [PubMed]
4. Xu, H.; Turlings, T.C.J. Plant volatiles as mate-finding cues for insects. *Trends Plant Sci.* **2018**, *23*, 100–111. [CrossRef] [PubMed]
5. Aldrich, J.R. Chemical communication in the true bugs and parasitoid exploitation. In *Chemical Ecology of Insects 2*; Ring, T.C., William, J.B., Eds.; Springer: Boston, MA, USA, 1995; pp. 318–363.
6. Vet, L.E.; Dicke, M. Ecology of infochemical use by natural enemies in a tritrophic context. *Annu. Rev. Entomol.* **1992**, *37*, 141–172. [CrossRef]
7. Tanaka, C.; Kainoh, Y.; Honda, H. Host frass as arrestant chemicals in locating host *Mythimna separata* by the tachinid fly *Exorista japonica*. *Entomol. Exp. Appl.* **2001**, *100*, 173–178. [CrossRef]
8. Ha, D.-M.; Choi, S.-H.; Shim, J.-K.; Jung, D.-O.; Song, K.-S.; Nho, S.-K.; Lee, K.-Y. Behavioral attraction of two parasitoids, *Venturia canescens* and *Bracon hebetor*, to silk extracts of a host *Plodia interpunctella*. *J. Asia-Pac. Entomol.* **2006**, *9*, 287–292. [CrossRef]
9. Filella, I.; Bosch, J.; Llusià, J.; Seco, R.; Peñuelas, J. The role of frass and cocoon volatiles in host location by *Monodontomerus aeneus*, a parasitoid of *Megachilid solitary* bees. *Environ. Entomol.* **2011**, *40*, 126–131. [CrossRef]
10. Agelopoulos, N.G.; Dicke, M.; Posthumus, M.A. Role of volatile inforchemicals emitted by feces of larvae in host-searching behavior of parasitoid *Cotesia rubecula* (Hymenoptera: Braconidae): A behavioral and chemical study. *J. Chem. Ecol.* **1995**, *21*, 1789–1811. [CrossRef]
11. Ngi-Song, A.J.; Overholt, W.A. Host location and acceptance by *Cotesia flavipes* Cameron and *C. sesamiae* (Cameron) (Hymenoptera: Braconidae), parasitoids of African gramineous stemborers: Role of frass and other host cues. *Biol. Control* **1997**, *9*, 136–142. [CrossRef]
12. Afsheen, S.; Wang, X.; Li, R.; Zhu, C.-S.; Lou, Y.-G. Differential attraction of parasitoids in relation to specificity of kairomones from herbivores and their by-products. *Insect Sci.* **2008**, *15*, 381–397. [CrossRef]
13. Dalen, M.; Knudsen, G.K.; Norli, H.R.; Thöming, G. Sources of volatiles mediating host location behaviour of *Glypta haesitator*, a larval parasitoid of *Cydia nigricana*. *Biol. Control* **2015**, *90*, 128–140. [CrossRef]
14. Sime, K.R. Experimental studies of the host-finding behavior of *Trogus pennator*, a parasitoid of swallowtail butterflies. *J. Chem. Ecol.* **2002**, *28*, 1377–1392. [CrossRef] [PubMed]
15. Weiss, M.R. Good housekeeping: Why do shelter-dwelling caterpillars fling their frass? *Ecol. Lett.* **2003**, *6*, 361–370. [CrossRef]
16. Weiss, M.R. Defecation behavior and ecology of insects. *Annu. Rev. Entomol.* **2006**, *51*, 635–661. [CrossRef]
17. Fatouros, N.E.; Huigens, M.E.; van Loon, J.J.A.; Dicke, M.; Hilker, M. Butterfly anti-aphrodisiac lures parasitic wasps. *Nature* **2005**, *433*, 704. [CrossRef] [PubMed]
18. Aldrich, J.R.; Khrimian, A.; Zhang, A.; Shearer, P.W. Bug pheromones (Hemiptera, Heteroptera) and tachinid fly host-finding. *Denisia* **2006**, *19*, 1015–1031.
19. Reddy, G.; Holopainen, J.; Guerrero, A. Olfactory responses of *Plutella xylostella* natural enemies to host pheromone, larval frass, and green leaf cabbage volatiles. *J. Chem. Ecol.* **2002**, *28*, 131–143. [CrossRef]
20. Obeysekara, P.T.; Legrand, A. The influence of host species and location in the host detection ability of Tiphiid (*Hymenoptera: Tiphiidae*) parasitoids. *Environ. Entomol.* **2014**, *43*, 1594–1602. [CrossRef]
21. Seenivasagan, T.; Chander, S.; Paul, A.N. Orientation and behavioural responses of *Cotesia plutellae* Kurdjumov (*Hymenoptera: Braconidae*) to cruciferous host plants and host larval body extracts. *J. Biol. Control* **2009**, *23*, 365–373.
22. Chiu-Alvarado, P.; Rojas, J.C. Host location behaviour by two *Cephalonomia* spp.(Hymenoptera: Bethylidae) wasps associated with the coffee berry borer *Hypothenemus hampei* (Coleoptera: Curculionidae: Scolytinae). *Int. J. Trop. Insect Sci.* **2008**, *28*, 179–184.
23. Jumean, Z.; Unruh, T.; Gries, R.; Gries, G. *Mastrus ridibundus* parasitoids eavesdrop on cocoon-spinning codling moth, *Cydia pomonella*, larvae. *Naturwissenschaften* **2005**, *92*, 20–25. [CrossRef] [PubMed]
24. Avila, G.A.; Withers, T.M.; Holwell, G.I. Olfactory cues used in host-habitat location and host location by the parasitoid *Cotesia urabae*. *Entomol. Exp. Appl.* **2016**, *158*, 202–209. [CrossRef]
25. Glasser, S.; Farzan, S. Host-associated volatiles attract parasitoids of a native solitary bee, *Osmia lignaria* Say (*Hymenoptera, megachilidae*). *J. Hymenopt. Res.* **2016**, *51*, 249. [CrossRef]
26. Rogers, M.E.; Potter, D.A. Kairomones from scarabaeid grubs and their frass as cues in below-ground host location by the parasitoids *Tiphia vernalis* and *Tiphia pygidialis*. *Entomol. Exp. Appl.* **2002**, *102*, 307–314. [CrossRef]
27. Steidle, J.L.; Steppuhn, A.; Ruther, J. Specific foraging kairomones used by a generalist parasitoid. *J. Chem. Ecol.* **2003**, *29*, 131–143. [CrossRef]
28. Steidle, J.L.; Fischer, A. Quantity does matter: How feces are used for host stage selection by granary weevil parasitoid *Lariophagus distinguendus*. *J. Chem. Ecol.* **2000**, *26*, 2657–2664. [CrossRef]
29. Meiners, T.; Westerhaus, C.; Hilker, M. Specificity of chemical cues used by a specialist egg parasitoid during host location. *Entomol. Exp. Appl.* **2000**, *95*, 151–159. [CrossRef]
30. Chuche, J.; Xuéreb, A.; Thiéry, D. Attraction of *Dibrachys cavus* (Hymenoptera: Pteromalidae) to its host frass volatiles. *J. Chem. Ecol.* **2006**, *32*, 2721–2731. [CrossRef]

31. Ramachandran, R.; Norris, D.M.; Phillips, J.K.; Phillips, T.W. Volatiles mediating plant-herbivore-natural enemy interactions: Soybean looper frass volatiles, 3-octanone and guaiacol, as kairomones for the parasitoid *Microplitis demolitor*. *J. Agric. Food Chem.* **1991**, *39*, 2310–2317. [CrossRef]
32. Röse, U.S.; Alborn, H.T.; Makranczy, G.; Lewis, W.J.; Tumlinson, J.H. Host recognition by the specialist endoparasitoid *Microplitis croceipes* (Hymenoptera: Braconidae): Role of host-and plant-related volatiles. *J. Insect Behav.* **1997**, *10*, 313–330. [CrossRef]
33. Mattiacci, L.; Hütter, E.; Dorn, S. Host location of *Hyssopus pallidus*, a larval parasitoid of the codling moth, *Cydia pomonella*. *Biol. Control* **1999**, *15*, 241–251. [CrossRef]
34. Weinhold, A.; Baldwin, I.T. Trichome-derived *O*-acyl sugars are a first meal for caterpillars that tags them for predation. *Proc. Natl. Acad. Sci. USA* **2011**, *108*, 7855–7859. [CrossRef] [PubMed]
35. Kang, W.-N.; Pan, Y.; Liao, L.-L.; Wu, Y.-K.; Zhang, X.-Q.; Jin, L.; Fu, K.-Y.; Guo, W.-C.; Li, G.-Q. Mandelonitrile produced by commensal bacteria protects the Colorado potato beetle against predation. *Nat. Commun.* **2024**, *15*, 10081. [CrossRef]
36. Wada-Katsumata, A.; Zurek, L.; Nalyanya, G.; Roelofs, W.L.; Zhang, A.; Schal, C. Gut bacteria mediate aggregation in the German cockroach. *Proc. Natl. Acad. Sci. USA* **2015**, *112*, 15678–15683. [CrossRef]
37. Desneux, N.; Ramírez-Romero, R.; Bokonon-Ganta, A.H.; Bernal, J.S. Attraction of the parasitoid *Cotesia marginiventris* to host (Spodoptera frugiperda) frass is affected by transgenic maize. *Ecotoxicology* **2010**, *19*, 1183–1192. [CrossRef]
38. Bourne, M.E.; Gloder, G.; Weldegergis, B.T.; Slingerland, M.; Ceribelli, A.; Crauwels, S.; Lievens, B.; Jacquemyn, H.; Dicke, M.; Poelman, E.H. Parasitism causes changes in caterpillar odours and associated bacterial communities with consequences for host-location by a hyperparasitoid. *PLoS Pathog.* **2023**, *19*, e1011262. [CrossRef]
39. Leroy, P.D.; Sabri, A.; Heuskin, S.; Thonart, P.; Lognay, G.; Verheggen, F.J.; Francis, F.; Brostaux, Y.; Felton, G.W.; Haubruge, E. Microorganisms from aphid honeydew attract and enhance the efficacy of natural enemies. *Nat. Commun.* **2011**, *2*, 348. [CrossRef]
40. Frago, E.; Zytynska, S. Impact of herbivore symbionts on parasitoid foraging behaviour. *Curr. Opin. Insect Sci.* **2023**, *57*, 101027. [CrossRef]
41. Cunningham, J.P.; Zalucki, M.P. Understanding *Heliothine* (Lepidoptera: Heliothinae) pests: What is a host plant? *J. Econ. Entomol.* **2014**, *107*, 881–896. [CrossRef]
42. Mlambo, S.; Mubayiwa, M.; Tarusikirwa, V.L.; Machekano, H.; Mvumi, B.M.; Nyamukondiwa, C. The fall armyworm and larger grain borer pest invasions in Africa: Drivers, impacts and implications for food systems. *Biology* **2024**, *13*, 160. [CrossRef]
43. Li, X.-J.; Wu, M.-F.; Ma, J.; Gao, B.-Y.; Wu, Q.-L.; Chen, A.-D.; Liu, J.; Jiang, Y.-Y.; Zhai, B.-P.; Early, R.; et al. Prediction of migratory routes of the invasive fall armyworm in eastern China using a trajectory analytical approach. *Pest Manag. Sci.* **2019**, *76*, 454–463. [CrossRef]
44. Sparks, A.N. A review of the biology of the fall armyworm. *Fla. Entomol.* **1979**, *62*, 82–87. [CrossRef]
45. Xu, S.; Hu, X.; Liu, Y.; Wang, X.; Wang, Y.; Li, G.; Turlings, T.C.; Li, Y. The threat of the fall armyworm to Asian rice production is amplified by the brown planthopper. *Plant Cell Environ.* **2025**, *48*, 1060–1072. [CrossRef] [PubMed]
46. Li, J.; Yan, F.; Coudron, T.A.; Pan, W.; Zhang, X.; Liu, X.; Zhang, Q. Field release of the parasitoid *Microplitis mediator* (Hymenoptera: Braconidae) for control of *Helicoverpa armigera* (Lepidoptera: Noctuidae) in cotton fields in Northwestern China's Xinjiang province. *Environ. Entomol.* **2006**, *35*, 694–699. [CrossRef]
47. Lu, Z.-y.; Yang, X.-f.; Ma, A.-h.; Ran, H.-f.; Liu, W.-x.; Li, J.-c. Parasitic potential of two *Microplitis* species on *Spodoptera frugiperda* larvae. *J. Hebei Agric. Sci.* **2020**, *24*, 37–39.
48. Arthur, A.P.; Mason, P.G. Life history and immature stages of the parasitoid *Microplitis mediator* (Hymenoptera: Braconidae), reared on the bertha armyworm *Mamestra configurata* (Lepidoptera: Noctuidae). *Can. Entomol.* **1986**, *118*, 487–491. [CrossRef]
49. Xu, H.; Desurmont, G.; Degen, T.; Zhou, G.; Laplanche, D.; Henryk, L.; Turlings, T. Combined use of herbivore-induced plant volatiles and sex pheromones for mate location in braconid parasitoids. *Plant Cell Environ.* **2017**, *40*, 330–339. [CrossRef]
50. Xu, H.; Veyrat, N.; Degen, T.; Turlings, T.C. Exceptional use of sex pheromones by parasitoids of the genus *Cotesia*: Males are strongly attracted to virgin females, but are no longer attracted to or even repelled by mated females. *Insects* **2014**, *5*, 499–512. [CrossRef]
51. Merkx-Jacques, M.; Bede, J.C. Influence of diet on the larval beet armyworm, *Spodoptera exigua*, glucose oxidase activity. *J. Insect Sci.* **2005**, *5*, 48. [CrossRef]
52. Xu, H.; Zhao, J.; Li, F.; Yan, Q.; Meng, L.; Li, B. Chemical polymorphism regulates the attractiveness to nymphs in the bean bug *Riptortus pedestris*. *J. Pest Sci.* **2021**, *94*, 463–472. [CrossRef]
53. Desurmont, G.A.; Xu, H.; Turlings, T.C. Powdery mildew suppresses herbivore-induced plant volatiles and interferes with parasitoid attraction in *Brassica rapa*. *Plant Cell Environ.* **2016**, *39*, 1920–1927. [CrossRef]
54. Xu, H.; Zhou, G.; Dötterl, S.; Schäffler, I.; Degen, T.; Chen, L.; Turlings, T.C.J. Distinct roles of cuticular aldehydes as pheromonal cues in two *Cotesia* parasitoids. *J. Chem. Ecol.* **2020**, *46*, 128–137. [CrossRef] [PubMed]
55. Steiner, S.; Mumm, R.; Ruther, J. Courtship pheromones in parasitic wasps: Comparison of bioactive and inactive hydrocarbon profiles by multivariate statistical methods. *J. Chem. Ecol.* **2007**, *33*, 825–838. [CrossRef] [PubMed]

56. Xu, H.; Wang, C.-W.; Zhang, R.; Liang, Y.; Li, M.-Q.; Ci, T.-T.; Wang, Y.-Y.; Wang, W.-J.; Zhang, Y.-N.; Pan, Z.-J.; et al. Harnessing Leguminosae typical scents for a more effective and eco-friendly pheromone trapping of pest. *J. Agric. Food Chem.* **2025**, *73*, 17519–17528. [CrossRef] [PubMed]
57. Bates, D.; Mächler, M.; Bolker, B.; Walker, S. lme4: Linear Mixed-Effects Models Using Eigen and S4. R Package Version 1.1–23. Available online: https://cran.r-project.org/web/packages/lme4 (accessed on 1 May 2025).
58. Xu, H.; Zhou, G.; Dötterl, S.; Schäffler, I.; von Arx, M.; Röder, G.; Degen, T.; Chen, L.; Turlings, T.C.J. The combined use of an attractive and a repellent sex pheromonal component by a gregarious parasitoid. *J. Chem. Ecol.* **2019**, *45*, 559–569. [CrossRef]
59. D'alessandro, M.; Erb, M.; Ton, J.; Brandenburg, A.; Karlen, D.; Zopfi, J.; Turlings, T.C. Volatiles produced by soil-borne endophytic bacteria increase plant pathogen resistance and affect tritrophic interactions. *Plant Cell Environ.* **2014**, *37*, 813–826. [CrossRef]
60. Lewis, W.; Tumlinson, J.H. Host detection by chemically mediated associative learning in a parasitic wasp. *Nature* **1988**, *331*, 257. [CrossRef]
61. Calatayud, P.-A.; Auger, J.; Thibout, E.; Rousset, S.; Caicedo, A.M.; Calatayud, S.; Buschmann, H.; Guillaud, J.; Mandon, N.; Bellotti, A.C. Identification and synthesis of a kairomone mediating host location by two parasitoid species of the cassava mealybug *Phenacoccus herreni*. *J. Chem. Ecol.* **2001**, *27*, 2203–2217. [CrossRef]
62. Chen, Y.; Ulyshen, M.D.; Poland, T.M. Abundance of volatile organic compounds in white ash phloem and emerald ash borer larval frass does not attract Tetrastichus planipennisi in a Y-tube olfactometer. *Insect Sci.* **2016**, *23*, 712–719. [CrossRef]
63. Virant-Doberlet, M.; Kuhelj, A.; Polajnar, J.; Šturm, R. Predator-prey interactions and eavesdropping in vibrational communication networks. *Front. Ecol. Evol.* **2019**, *7*, 203. [CrossRef]
64. Cortesero, A.; De Moraes, C.M.; Stapel, J.; Tumlinson, J.; Lewis, W. Comparisons and contrasts in host-foraging strategies of two larval parasitoids with different degrees of host specificity. *J. Chem. Ecol.* **1997**, *23*, 1589–1606. [CrossRef]
65. Sidorov, R.; Zhukov, A.; Pchelkin, V.; Tsydendambaev, V. Palmitic acid in higher plant lipids. In *Palmitic Acid: Occurrence, Biochemistry and Health Effects*; NOVA Publishers: New York, NY, USA, 2014; pp. 125–144.
66. Tonelli, M.; Gomes, G.; Silva, W.D.; Magri, N.T.; Vieira, D.M.; Aguiar, C.L.; Bento, J.M.S. Spittlebugs produce foam as a thermoregulatory adaptation. *Sci. Rep.* **2018**, *8*, 4729. [CrossRef] [PubMed]
67. Noushini, S.; Perez, J.; Park, S.J.; Holgate, D.; Jamie, I.; Jamie, J.; Taylor, P. Rectal gland chemistry, volatile emissions, and antennal responses of male and female banana fruit fly, *Bactrocera musae*. *Insects* **2020**, *11*, 32. [CrossRef]
68. Evans, P.H.; Becerra, J.X.; Venable, D.L.; Bowers, W.S. Chemical analysis of squirt-gun defense in *Bursera* and counter defense by *Chrysomelid* beetles. *J. Chem. Ecol.* **2000**, *26*, 745–754. [CrossRef]
69. Jing, T.-Z.; Qi, F.-H.; Wang, Z.-Y. Most dominant roles of insect gut bacteria: Digestion, detoxification, or essential nutrient provision? *Microbiome* **2020**, *8*, 38. [CrossRef]
70. Weisskopf, L.; Schulz, S.; Garbeva, P. Microbial volatile organic compounds in intra-kingdom and inter-kingdom interactions. *Nat. Rev. Microbiol.* **2021**, *19*, 391–404. [CrossRef]
71. Basit, A.; Haq, I.U.; Hyder, M.; Humza, M.; Younas, M.; Akhtar, M.R.; Ghafar, M.A.; Liu, T.-X.; Hou, Y. Microbial symbiosis in Lepidoptera: Analyzing the gut microbiota for sustainable pest management. *Biology* **2025**, *14*, 937. [CrossRef]

MDPI AG
Grosspeteranlage 5
4052 Basel
Switzerland
Tel.: +41 61 683 77 34

Biology Editorial Office
E-mail: biology@mdpi.com
www.mdpi.com/journal/biology